NUMERICAL SIMULATION IN HYDRAULIC FRACTURING:
MULTIPHYSICS THEORY AND APPLICATIONS

Multiphysics Modeling

Series Editors

Jochen Bundschuh
University of Southern Queensland (USQ), Toowoomba, Australia
Royal Institute of Technology (KTH), Stockholm, Sweden

Mario César Suárez Arriaga
Private Consultant, Thermomechanics of Porous Rocks, Morelia,
Michoacán, Mexico

ISSN: 1877-0274

Volume 8

Numerical Simulation in Hydraulic Fracturing: Multiphysics Theory and Applications

Xinpu Shen

Guoyang Technology and Services, Houston, TX, USA

William Standifird

Halliburton, Consulting, Houston, TX, USA

CRC Press
Taylor & Francis Group
Boca Raton London New York

CRC Press is an imprint of the
Taylor & Francis Group, an **informa** business

A BALKEMA BOOK

Published by:
CRC Press/Balkema
P.O. Box 447, 2300 AK Leiden, The Netherlands
e-mail: Pub.NL@taylorandfrancis.com
www.crcpress.com – www.taylorandfrancis.com

First issued in paperback 2020

ISBN 13: 978-0-367-57381-2 (pbk)
ISBN 13: 978-1-138-02962-0 (hbk)

Visit the Taylor & Francis Web site at
http://www.taylorandfrancis.com

and the CRC Press Web site at
http://www.crcpress.com

Typeset by MPS Limited, Chennai, India

Library of Congress Cataloging-in-Publication Data

About the book series

Numerical modeling is the process of obtaining approximate solutions to problems of scientific and/or engineering interest. The book series addresses novel mathematical and numerical techniques with an interdisciplinary emphasis that cuts across all fields of science, engineering and technology. It focuses on breakthrough research in a richly varied range of applications in physical, chemical, biological, geoscientific, medical and other fields in response to the explosively growing interest in numerical modeling in general and its expansion to ever more sophisticated physics. The goal of this series is to bridge the knowledge gap among engineers, scientists, and software developers trained in a variety of disciplines and to improve knowledge transfer among these groups involved in research, development and/or education.

This book series offers a unique collection of worked problems in different fields of engineering and applied mathematics and science, with a welcome emphasis on coupling techniques. The book series satisfies the need for up-to-date information on numerical modeling. Faster computers and newly developed or improved numerical methods such as boundary element and meshless methods or genetic codes have made numerical modeling the most efficient state-of-the-art tool for integrating scientific and technological knowledge in the description of phenomena and processes in engineered and natural systems. In general, these challenging problems are fundamentally coupled processes that involve dynamically evolving fluid flow, mass transport, heat transfer, deformation of solids, and chemical and biological reactions.

This series provides an understanding of complicated coupled phenomena and processes, its forecasting, and approaches in problem solving for a diverse group of applications, including natural resources exploration and exploitation (e.g. water resources and geothermal and petroleum reservoirs), natural disaster risk reduction (earthquakes, volcanic eruptions, tsunamis), evaluation and mitigation of human induced phenomena (climate change), and optimization of engineering systems (e.g. construction design, manufacturing processes).

Jochen Bundschuh
Mario César Suárez Arriaga
(Series Editors)

Editorial board

Table of Contents

Foreword by M.Y. Soliman

Hydraulic fracturing had impacted the oil industry probably more than any other technology. Starting in the late 1940's, Hydraulic fracturing helped offset the declining production from lower permeability formation by providing a pathway for reservoir fluids to flow towards the wellbore. Hydraulic fracturing became especially important as the price of crude oil and natural gas increased and the industry turned to lower permeability reservoirs to satisfy the steadily increasing global demand for energy. The 1970's and the early 1980's saw a spike in hydraulic fracturing activities. "Massive Hydraulic Fracturing" became a term that described the creation of a large hydraulic fracture intersecting a vertical well located in a low permeability formation. Massive hydraulic fracturing were usually several hundreds of feet long and involved the injection of hundreds of thousands of pounds of proppant, and in some cases more than a million pounds of proppant was injected during the treatment. Massive Hydraulic Fracturing allowed the economic exploitation of gas reservoirs that would have been otherwise uneconomical to produce.

Starting the late 1980's, the combination of hydraulic fracturing with the new advancement of drilling horizontal wells added a new dimension to the industry that eventually opened the door to production from formation with ultra-low permeability such as shale. Shale, CBM, and ultra-low permeability formations are commonly referred to as unconventional reservoirs. The combination of hydraulic fracturing and horizontal well drilling created a new production opportunity, however it also created significant technical challenges, especially in the geomechanical aspect of fracturing. The interference between the created hydraulic fractures both in fluid flow and stresses and the existing natural fractures are areas of great technical challenge and opportunity. As we have seen recently in industrial applications some techniques use this interference (stress shadowing) between fractures to enhance the effect of hydraulic fracturing on production. Alternating fracturing, zipper-frac, and modified zipper-frac are techniques used exactly for that purpose. Other challenges include seismicity, casing integrity, and frac bashing. Another very significant challenge is re-frac of horizontal wells previously fractured to exploit reservoir volumes that have not been properly drained of hydrocarbon.

The degree of science and engineering in working with these new technical challenges could be overwhelming. The best approach to work with those challenges is to use a numerical simulator to study those challenges and design the fracturing treatment to avoid potential problems and enhance the performance of the fracturing treatment. This book thoroughly investigates the use of a numerical simulator to address many of the issues that face an engineer studying the geomechanical issues of fracturing vertical and horizontal wells. It fills a gap that exist in literature. Other issues that are addressed in the book is the issue of wellbore stability and casing integrity. Both issues are of paramount importance in drilling and fracturing horizontal wells.

M. Y. Soliman, PhD, PE, NAI
Chair of the Petroleum Engineering Department
William C Miller Endowed Chair
The University of Houston

Authors' preface

The expansion of unconventional petroleum resources in the recent decade and the rapid development of computational technology have provided the opportunity to develop and apply 3D numerical modeling technology to simulate the hydraulic fracturing of shale and tight sand formations. This book presents 3D numerical modeling technologies for hydraulic fracturing developed in recent years, and introduces solutions to various 3D geomechanical problems related to hydraulic fracturing. In the solution processes of the case studies included in the book, fully coupled multi-physics modeling has been adopted, along with innovative computational techniques, such as submodeling.

In practice, hydraulic fracturing is an essential project component in shale gas/oil development and tight sand oil, and provides an essential measure in the process of drilling cuttings reinjection (CRI). It is also an essential measure for widened mud weight window (MWW) when drilling through naturally fractured formations; the process of hydraulic plugging is a typical application of hydraulic fracturing.

3D modeling and numerical analysis of hydraulic fracturing is essential for the successful development of tight oil/gas formations: it provides accurate solutions for optimized stage intervals in a multistage fracking job. It also provides optimized well-spacing for the design of zipper-frac wells.

Quasi-brittle fracturing is one of the fundamental mechanisms in the hydraulic fracturing process. Continuum damage mechanics (CDM) is one of the theoretical tools used for the 3D description of the initiation and propagation of fractures, both for natural fractures and for fractures induced by hydraulic injection. In this book, a CDM-based method is used to optimize value of stage-intervals for multistage fracturing, and value of well-spacing for a pair of zipper-frac wells. CDM-based 3D modeling can simulate the development of fracture clouds in a 3D volume. Consequently, it is a real-3D tool, as compared to those models that use 3D-planar fractures. Cohesive element is also used to simulate the CRI process and to investigate the widened MWW.

Numerical estimation of casing integrity under stimulation injection in the hydraulic fracturing process is one of major concerns in the successful development of unconventional resources. This topic is also investigated numerically in this book. Numerical solutions to several other typical geomechanics problems related to hydraulic fracturing, such as fluid migration caused by fault reactivation and seismic activities, are also presented.

Although examples presented here are restricted essentially to applications in petroleum industry, but methods and basic techniques introduced in the book can also be used in geothermal engineering which essentially works into volcanic rocks.

We hope the contents of this book are helpful to readers on their way to achieving higher levels of their careers.

Xinpu Shen
William Standifird
January 2017

About the authors

Xinpu Shen is a Senior Advisor at Guoyang Technology and Services, and formerly a Senior Advisor at Halliburton Consulting.

He received his PhD degree in Engineering Mechanics in 1994 from Tsinghua University, Beijing, China. He was lecturer and associate professor in Tsinghua University from 1993 to 1999. Since May 2001, he was a professor in Engineering Mechanics in Shenyang University of Technology, China. From 1997 to 2004, he worked as postdoctoral research associate in several European institutions, including Politecnico di Milano, Italy and the University of Sheffield, UK, etc. He worked as consultant of geomechanics for Knowledge Systems Inc Houston since 2005 and until it was acquired by Halliburton in 2008. He has been coordinator to 4 projects supported by the National Natural Science Foundation of China since 2005. He is inventor (co-inventor) to 10 patents and author (coauthor) to 7 books and 98 related papers among which 30 can be downloaded from OnePetrol.com.

William Standifird currently serves as a Director – Global Technical Practices at Halliburton. In this role he is charged with the invention, development and deployment of innovative technologies that support safe and efficient well construction for petroleum assets. William began his career with Schlumberger as a Drilling Services Engineer where he specialized in the application of petroleum geomechanics to deepwater drilling operations. He subsequently joined Knowledge Systems Inc. and rapidly built a global petroleum geomechanics practice which was acquired by Halliburton in 2008. William has over 20 peer reviewed publications, a Performed by Schlumberger Silver Medal and a Hart's Meritorious Engineering Award. He holds undergraduate degrees in electronics engineering, management science and earned a Master of Business Administration from the University of Houston System.

Acknowledgements

Warmest thanks are due to Professor Jochen Bundschuh and Professor Mario Suárez, Editor and co-editor of the book series in which this book is one volume, for many very helpful academic communications from them.

Thanks are due to Dr Xiaomin Hu and Mr Guoyang Shen for their help in preparing User Subroutines and data processing tools used during numerical modeling.

Thanks are due to Professor Kaspar Willam from Houston University, Dr Jie Bai from Production Enhancement PSL of Halliburton, Dr Jiachun Wang and Dr Bo Wang from the American Bureau of Shipping, Dr Yaping Zhu from Statoil, for their kind help during the preparation of this book.

Partial Financial support from NSF of China through contract (11272216) to North China University of Technology, is gratefully acknowledged.

Xinpu Shen
William Standifird
January, 2017

CHAPTER 1

Introduction to continuum damage mechanics for rock-like materials

Continuum damage mechanics was a part of fracture mechanics at its early stage, but later developed into an independent branch within the framework of solid mechanics. With fracture mechanics, it jointly forms failure mechanics in a wide range of academic domains.

This chapter introduces the fundamental concept and principles of continuum damage mechanics. This introduction is followed by the typical models used in the analysis of geomechanics. The numerical scheme for the calculation of the damage model is also introduced in detail through a typical example.

1.1 INTRODUCTION

Rock-like materials have a complex mechanical behavior, such as anisotropy, hysteresis, dilatancy, irreversible strain, and strongly path-dependent stress-strain relationships. The nonlinear behavior of rock-like materials is generally associated with the existence of many cracks and their propagation.

Micromechanical approaches attempted to predict the macroscale thermomechanical response of heterogeneous materials, based on mesostructural models of a representative volume element (RVE) within the material. Micromechanical models have the distinct advantages of being able to capture structure details at the microscale and mesoscale, and to enable the formulation of the kinetic equation for damage evolution based on the physical process involved. These models, however, can be computationally inefficient in many practical applications and can therefore be applied only to limited cases.

Continuum damage mechanics (CDM) uses continuum variables to describe cracks and joints, thus providing another means of modeling a jointed rock mass. CDM (Lemaitre, 1990) is based on the thermodynamics of irreversible process, internal state variable theory, and relevant physical considerations. Formal CDM modeling of the cracked rock was first suggested by Kawamoto *et al.* (1988). Thereafter, additional advances were reported in simple cases of anisotropic damage analysis. Despite all of the efforts described, it has only now become possible to directly use anisotropic CDM theory for a jointed rock mass mechanical analysis.

The inelastic failure of rock-like materials and structures is characterized by the initiation and evolution of cracks and the frictional sliding on the closed-crack surfaces. Plastic damage models are the major measures used to address with cracking-related failure analysis and are widely used by various researchers (Chaboche, 1992; de Borst *et al.*, 1999; Lemaitre, 1990; Seweryn and Mroz, 1998). At room temperature, the mechanical behavior of concrete, as a typical rock-like material, is similar to that of rock. Its damage evolution behavior has been investigated widely. Several damage models for concrete-like materials are reported in the references.

One of the popular models used in practice is the Mazars-Pijaudier-Cabot damage model (Mazars and Pijaudier-Cabot, 1989). Because this damage model is expressed in a holonomic manner with respect to strain loading, no incremental form was presented for the damage evolution law. Consequently, the damage evolution is uncoupled with the evolution of inelastic strain.

The de Borst *et al.* (1999) damage model is another popular model. In this damage model, the damage evolution law is based on the total quantity of equivalent strain. In practice, a great deal of strength data are available in terms of equivalent stress and/or fracture energy, but existing data are lacking for strain-based criteria. For this reason, the equivalent-strain-based damage model

1

will require special tests. Consequently, for this consideration, a plasticity-based damage model with equivalent-stress-based loading criterion is more practical than strain-based models.

The Barcelona model is one of the plasticity-based damage models; it is reported by Lubliner *et al.* (1989) and adopted by Lee and Fenves (1998) and Nechnech *et al.* (2002). In this model, a holonomic relationship between damage and equivalent plastic strain is presented. Two damage variables are designed: one for tensile damage and one for compressive damage.

An anisotropic damage model of the fourth-order tensor form is reported by Govindjee *et al.* (1995) for the simulation of concrete plastic damage. This model has been later extended by Meschke *et al.* (1998) to Drucker-Prager type plasticity and plastic damage. The advantages of this model include the following:

- Because the damage is implicitly consisted in its elastic compliance tensor, there is no need to adopt the effective stress concept in this model.
- The inelastic calculation could be performed in nominal stress space.
- There is no limit for any kind of anisotropy.

However, Meschke *et al.* (1998) showed that some weakness points exist in the model with respect to its energy dissipation property. In addition, the proportion of plastic strain with respect to damage strain was designed to be governed by a constant parameter. Another disadvantage of the model is that it was only used for the analysis of Mode I (smeared-crack) fracture problems (because the Mode I fracture energy was adopted in the previously mentioned references). Furthermore, it is actually difficult to connect the stress-triaxiality dependent softening-phenomena with any kind of fracture modes.

Anisotropic damage models of vector form and of second-order tensor form were also investigated by several researchers (Swoboda *et al.*, 1998). However, because of the difficulties in the simulation of the experimental phenomena of rock-like specimen, as well as its inconvenience on computational aspects, it is still not appropriate to adopt these models in this study.

Because of its simplicity and a reasonable capacity of problem representation, the isotropic damage model is the most popular damage model in the simulation of the failure phenomena of rock-like materials and structures. Therefore, it is the choice of this work, despite that the damage of rock usually appears in the form of directional cracks and that the plasticity of concrete is often related to the confined frictional sliding between closed-crack surfaces under room temperature. Anisotropic phenomena will be simulated by the scale damage model on the structural level, rather than on the material level.

Only damage models of scalar form will be introduced in this chapter. The introduction of more complex damage models in vector form and tensor form for rock can be found in Swoboda *et al.* (1998). The contents of this chapter include the following:

- Section 1.2 introduces the incremental form of damage model developed by Lubliner *et al.* (1996) and Lee and Fenves (1998).
- Section 1.3 introduces the holonomic form of damage model developed by Mazars *et al.* (2014).
- Section 1.4 introduces the process for the development of a Drucker-Prager type plastic damage model; the time-integration numerical-scheme of the proposed constitutive model is introduced in detail. The validation process of the plastic damage model at a local level is also presented, with examples of various loading cases.
- Concluding remarks are provided at the end of the chapter.

1.2   THE BARCELONA MODEL: SCALAR DAMAGE WITH DIFFERENT BEHAVIORS FOR TENSION AND COMPRESSION

The Barcelona model was first proposed by Lubliner *et al.* (1996) and further improved by Lee and Fenves (1998). It is a plasticity-based scalar continuum damage model. The mechanism for

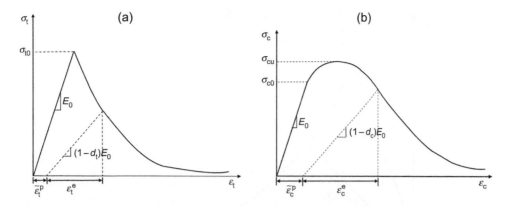

Figure 1.1   Uniaxial behavior of the model (a) under tension and (b) under compression.

damage evolution in this model includes two aspects: damage resulting from tensile cracking and damage resulting from compressive crushing. The evolution of plastic loading is controlled by two hardening parameters: the equivalent plastic strain ($\overline{\varepsilon}_t^p$), which is caused by tensile load, and the part ($\overline{\varepsilon}_t^p$), which is caused by compressive load.

1.2.1   Uniaxial behavior of the Barcelona model

As shown in Figure 1.1a, the material indicates linear elastic behavior before the stress reaches the value of σ_{t0}. Microcracking/damage initiates as the stress values exceeds the point of σ_{t0}. The strain-softening phenomenon appears as a result of damage evolution, and will result in strain localization. The term 'microcrack' used here refers to those very small cracks at where material is not yet completely broken. There are still connections of material at macro scope at the place of micro cracks. Because there is no specific usage of microcrack here, it is just used as a concept for similarity to the concept of continuum damage.

For the compressive behavior shown in Figure 1.1b, the material also indicates linear elastic behavior before the stress reaches the value of σ_{c0}. Microcracking/damage begins as the stress values exceed σ_{c0}. Strain-hardening occurs and will persist until the stress level reaches σ_{cu}. As the stress level exceeds σ_{cu}, the strain-softening phenomenon appears as a result of damage evolution, and will result in strain localization.

Because of damage initiation and evolution, unloading stiffness is degraded from its original value of the intact material, as shown in Figure 1.1. This stiffness degradation is expressed in terms of two damage variables: d_t and d_c. The value of the damage variables are 0 for the intact material and 1 for the completely broken state.

Assuming that E_0 is the Young's modulus value of the initial intact material, the Hooke's law under uniaxial loading conditions is:

$$\sigma_t = (1 - d_t)E_0(\varepsilon_t - \overline{\varepsilon}_t^p) \tag{1.1}$$

$$\sigma_c = (1 - d_c)E_0(\varepsilon_c - \overline{\varepsilon}_c^p) \tag{1.2}$$

Therefore, the effective stress for tension and compression can be written as:

$$\overline{\sigma}_t = \frac{\sigma_t}{1 - d_t} = E_0(\varepsilon_t - \overline{\varepsilon}_t^p) \tag{1.3}$$

$$\overline{\sigma}_c = \frac{\sigma_c}{1 - d_c} = E_0(\varepsilon_c - \overline{\varepsilon}_c^p) \tag{1.4}$$

Plastic yielding criteria will be described in the space of effective stress.

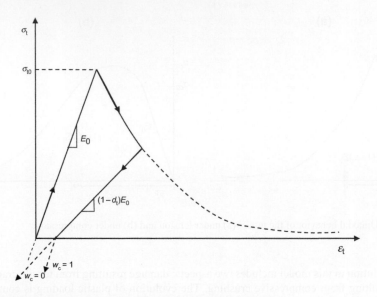

Figure 1.2   Illustration of the unloading behavior of the Barcelona model.

1.2.2 Unloading behavior

A description of the unloading behavior of the model is important when applying the model to periodic loading conditions. Closures and openings of the existing microcracks will result in significant nonlinearity in the behavior of the material. Experiments have proven that crack closure will result in some degree of stiffness recovery, which is also known as the "unilateral effect."

The relationship of Young's modulus (E) for damaged material and that of the intact material (E_0) is:

$$E = (1 - D)E_0 \tag{1.5}$$

Lemaitre's "strain equivalent assumption" is adopted. In Equation (1.5), D is the synthetic damage variable, which is a function of stress state (s), tensile damage (d_t), and compressive damage (d_c), and can be expressed as:

$$(1 - D) = (1 - s_t d_t)(1 - s_c d_c) \tag{1.6}$$

where s_t and s_c are functions of the stress state and, are calculated as follows:

$$s_t = 1 - w_t \gamma^*(\sigma_{11}); \quad 0 \le w_t \le 1 \tag{1.7}$$

$$s_c = 1 - w_c(1 - \gamma^*(\sigma_{11})); \quad 0 \le w_c \le 1 \tag{1.8}$$

In Equation (1.7), $\gamma^*(\sigma_{11}) = H(\sigma_{11}) = \begin{cases} 1 & \text{as } \sigma_{11} > 0 \\ 0 & \text{as } \sigma_{11} \le 0 \end{cases}$

The weight parameters, w_t and w_c, are the material properties that control the amount of stiffness recovery for the unloading process, as shown in Figure 1.2.

Hooke's law at the triaxial stress state is expressed in the following tensor form:

$$\boldsymbol{\sigma} = (1 - D)\mathbf{E}_0^e : (\boldsymbol{\varepsilon} - \boldsymbol{\varepsilon}^p) \tag{1.9}$$

where $\mathbf{E}_0^e$ is the matrix stiffness of the intact material. At a triaxial stress state, the Heaviside function γ in the expression of the synthetic damage variable d can be written as:

$$\gamma(\hat{\sigma}) = \frac{\sum_{i=1}^{3}(\sigma_i)}{\sum_{i=1}^{3}|\hat{\sigma}_i|}; \quad 0 \le \gamma(\hat{\sigma}) \le 1 \tag{1.10}$$

where $\hat{\sigma}_i$ $(i = 1, 2, 3)$ are the principal stress components, and the symbol $< \cdot >$ indicates that $\langle x \rangle \geq \frac{1}{2}(|x| + x)$.

1.2.3 Plastic flow

The non-associated plastic flow rule is adopted in the model. The plastic potential G_p is in the form of a Drucker-Prager type and is expressed as:

$$G_p = \sqrt{(\varepsilon \sigma_{t0} tg \Psi)^2 + \overline{q}^2} - \overline{p} tg \Psi \tag{1.11}$$

where $\Psi(\theta, f_i)$ is the dilatancy angle, and $\sigma_{t0}(\theta, f_i) = \sigma_t|_{\overline{\varepsilon}^{pl}=0}$ is the threshold value of the tensile stress at which damage initiates. Parameter $\varepsilon(\theta, f_i)$ is a model parameter that defines the eccentricity of the loading surface in the effective stress space.

1.2.4 Yielding criterion

The yielding criterion of the model is given in the effective stress space, and its evolution is determined by two variables, $\overline{\varepsilon}_t^p$ and $\overline{\varepsilon}_c^p$. Its expression is provided in:

$$f = \frac{1}{1 - \alpha}(\overline{q} - 3a_1\overline{p} + a_2(\overline{\varepsilon}^p)\langle \hat{\overline{\sigma}}_{max} \rangle - a_3 \langle -\hat{\overline{\sigma}}_{max} \rangle) - \overline{\sigma}_c(\overline{\varepsilon}_c^p) = 0 \tag{1.12}$$

where:

$$a_1 = \frac{\left(\frac{\sigma_{b0}}{\sigma_{c0}}\right) - 1}{2\left(\frac{\sigma_{b0}}{\sigma_{c0}}\right) - 1}, \quad 0 \leq a_1 \leq 0.5, \quad a_2 = \frac{\overline{\sigma}_c(\overline{\varepsilon}_c^p)}{\overline{\sigma}_t(\overline{\varepsilon}_t^p)}(1 - a_1) - (1 + a_1),$$

$$a_3 = \frac{3(1 - k_c)}{2k_c - 1}$$

where $\hat{\overline{\sigma}}_{max}$ is the value of the maximum principal stress component; σ_{b0}/σ_{c0} is the ratio between the limit of bi-axial and uniaxial compression. Parameter k_c is the ratio between the second invariant of the stress tensor at the tensile meridian $q(TM)$ and the second invariant at the compressive meridian $q(CM)$ under arbitrary pressure, which makes $\hat{\sigma}_{max} > 0$. It is required that $0.5 < k_c \leq 1.0$ have a default value of 2/3. In Equation (1.12), $\overline{\sigma}_t(\overline{\varepsilon}_t^p)$ is the effective tensile strength, and $\overline{\sigma}_c(\overline{\varepsilon}_c^p)$ is the effective compressive strength. Figure 1.3 shows the yielding surface for various values of k_c at π-plane. Figure 1.4 shows the yielding surface for the plane-stress state.

1.3 MAZARS'S HOLONOMIC FORM OF CONTINUUM DAMAGE MODEL

1.3.1 Concepts

In the Mazars' holonomic damage model (Mozars *et al.*, 2014), two equivalent strain variables are defined with ε_t for tensile strain and ε_c for compressive strain, and are calculated in Equation (1.13):

$$\varepsilon_t = \frac{I_\varepsilon}{2(1 - 2\nu)} + \frac{\sqrt{J_\varepsilon}}{2(1 + \nu)}, \quad \varepsilon_c = \frac{I_\varepsilon}{5(1 - 2\nu)} + \frac{6\sqrt{J_\varepsilon}}{5(1 + \nu)} \tag{1.13}$$

where ν is the Poisson's ratio, I_ε is the first invariant of the strain tensor, and J_ε is the second invariant of the deviatoric strain tensor.

Lemaitre's strain equivalent principle is adopted for the relationship between Young's modulus and damage variable d. For the case of uniaxial loading, it is expressed as:

$$E = E_0(1 - D) \tag{1.14}$$

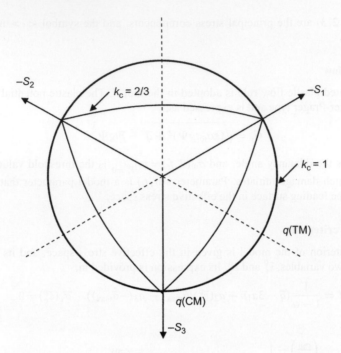

Figure 1.3   The yielding surface for various values of k_c at π-plane.

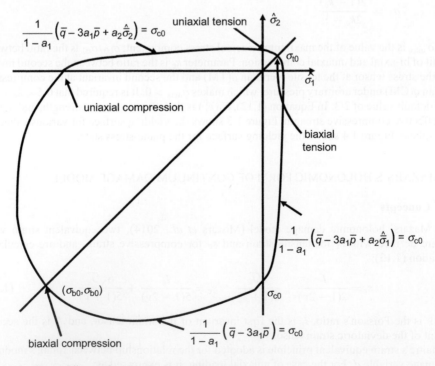

Figure 1.4   The yielding surface for the plane stress state.

where D represents the continuum damage variable, E_0 represents the initial value of Young's modulus of intact material, and E represents the degraded value of Young's modulus of damaged material.

1.3.2 Criterion of damage initiation

The Mazars' damage model uses the following criteria for damage initiation:

$$f_t = \varepsilon_t - Y_t \leq 0 \tag{1.15}$$

$$f_c = \varepsilon_c - Y_c \leq 0 \tag{1.16}$$

where Y_t and Y_c are parts of the thermodynamic conjugate force Y of the damage variable D for the case of loading in tension and compression, respectively. Their definitions are given in Equation (1.17) and Equation (1.18):

$$Y_t = \max(\varepsilon_{t0}, \max(\varepsilon_t)) \tag{1.17}$$

$$Y_c = \max(\varepsilon_{c0}, \max(\varepsilon_c)) \tag{1.18}$$

In Equation (1.17) and Equation (1.18), ε_{t0} and ε_{c0} are thresholds for damage initiation during loading for the case of tension and compression, respectively.

The definition of thermodynamic conjugate force Y is given as:

$$Y = rY_t + (1 - r)Y_c \tag{1.19}$$

where r is the triaxiality factor (Lee and Fenves, 1980); it is equal to 1 for tensile loading and 0 for compression loading.

1.3.3 Damage evolution law

The damage evolution law given in Mazars' holonomic damage model is in the form of total quantity, rather than incremental:

$$D = 1 - \frac{(1 - A)Y_0}{Y} - Ae^{-B(Y - Y_0)} \tag{1.20}$$

where Y_0 is the initial threshold of the damage-conjugate force Y with given values of ε_{t0} and ε_{c0}, respectively. A and B are two loading factors; they are functions of the triaxiality r, and are defined as:

$$A = A_t[2r^2(1 - 2k) - r(1 - 4k)] + A_c(2r^2 - 3r + 1) \tag{1.21}$$

$$B = B_t r^{r^2 - 2r + 2} + B_c(1 - r^{r^2 - 2r + 2}) \tag{1.22}$$

where A_t, A_c, B_t, and B_c are all material parameters; k is the ratio of A/A_t and is used to calibrate pure shear behavior (Mazars *et al.*, 2013). The values of these material parameters can be calibrated with experimental data, as described by Mazars *et al.* (2013).

The major advantage of the Mazars' holonomic damage model is that the value of the damage variable can be calculated for a given strain status, and an iteration at the local level of a material point is not needed. The major disadvantage is there are many material parameters, which requires calibrated with experimental data.

1.4   SUBROUTINE FOR UMAT AND A PLASTIC DAMAGE MODEL WITH STRESS TRIAXIALITY-DEPENDENT HARDENING

1.4.1   **Introduction**

The subroutine for user-developed material model (UMAT) is a popular method used by engineers to implement a nonlinear material model into large-scale commercial finite element method (FEM) software. This section introduces the procedure for developing a UMAT and validating its accuracy at a local level of a material point.

The content of this subsection is organized in the following order:

- Section 1.4.2 provides the equations of the constitutive model for elasto-plasticity coupled with damage in general; a definition of stress triaxiality is proposed and later introduced in the plastic hardening and damage evolution laws.
- Section 1.4.3 describes the development of a driver subroutine for the validation of the constitutive models with reference to the principle proposed by Hashash *et al.* (2002). The results of numerical tests of the proposed model are provided for typical loading cases.
- Section 1.4.4 includes concluding remarks.

1.4.2   **Formulation of the proposed model**

1.4.2.1   *Fundamental equations of the plasticity-based damage model*

With the "Energy Equivalence Principle," the fundamental relationships of the plasticity-based damage model proposed by Saanouni *et al.* (1994) are listed in Equation (1.23):

$$
\begin{cases}
\tilde{\sigma}_{ij} = \dfrac{\sigma_{ij}}{1-D}, \quad \tilde{\boldsymbol{\varepsilon}}^{\mathrm{e}}_{ij} = (1-D)\boldsymbol{\varepsilon}^{\mathrm{e}}_{ij}, \quad \tilde{\mathbf{E}}^{0}_{ijkl} = \dfrac{\mathbf{E}_{ijkl}}{(1-D)^2} \\[2mm]
\tilde{\mathbf{I}}_1 = \tilde{\sigma}_{ii}, \quad \tilde{J}_2 = \dfrac{1}{2}\tilde{\mathbf{s}}_{ij}\tilde{\mathbf{s}}_{ij}, \quad \tilde{\mathbf{s}}_{ij} = \tilde{\sigma}_{ij} - \dfrac{\tilde{\mathbf{I}}_1}{3} \\[2mm]
Y = (1-D)\mathbf{E}^{0}_{ijkl}\boldsymbol{\varepsilon}^{\mathrm{e}}_{ij}\boldsymbol{\varepsilon}^{\mathrm{e}}_{kl} \\[2mm]
\sigma_{ij} = \mathbf{E}^{0}_{ijkl}(1-D)^2(\boldsymbol{\varepsilon}_{kl} - \boldsymbol{\varepsilon}^{\mathrm{P}}_{kl}) \\[2mm]
\dot{\boldsymbol{\varepsilon}}^{\mathrm{p}}_{ij} = \dot{\lambda}\dfrac{\partial F}{\partial \sigma_{ij}}, \quad \dot{D} = \dot{\lambda}\dfrac{\partial F}{\partial Y}
\end{cases}
\tag{1.23}
$$

where σ_{ij} represents the total stress tensor; ε_{ij} represents the total strain tensors; superscript p represents plastic quantities; e represents elastic quantities; tilde ($\sim$) represents quantities for fictitious net materials; $\tilde{\mathbf{s}}_{ij}$ represents the deviatoric stress tensor; λ represents the inelastic multiplier; D represents the isotropic damage variable; Y represents damage conjugate force; $\mathbf{E}^{0}_{ijkl}$ represents the elasticity tensor of the intact material; δ_{ij} represents a second-order unit tensor; $\tilde{\mathbf{I}}_1$ represents the sum of the effective principal stresses; $\tilde{J}_2$ represents a second invariant of the deviatoric effective stress tensor; F represents a plastic damage potential function; and $\tilde{Q}$ represents the plastic part of the potential in the effective stress space.

These equations show that the damage evolution is designed to be accompanied by a plastic strain increase, and its quantity of increment also depends on the elastic-strain tensor by means of the damage-conjugate force.

1.4.2.2   *Specification for Drucker-Prager type plasticity coupled with damage*

With reference to the generalized Drucker-Prager criterion introduced in Menétrey and Willam (1995) and the hardening model introduced in Besson (2001), the plastic damage loading condition is primarily defined in the effective stress space in the following form (stress triaxiality is introduced in a later section):

$$
\tilde{f} = \alpha_{\mathrm{F}}\tilde{I}_1 + \tilde{J}_2^{1/2} - [k + k_\infty(1 - \mathrm{e}^{-b\lambda})] = 0
\tag{1.24}
$$

where k represents the initial shear-strength constant; k_∞ represents the strain hardening limit of the fictitious net material, which corresponds to infinite equivalent plastic strain, i.e. $\lambda \to \infty$; α_F represents a material constant designed for pressure-sensitivity properties; b represents a model constant, which can be determined by fitting the experimental phenomena.

The plastic part of the potential (i.e., $\tilde{Q}$) is given in the effective stress space as:

$$\tilde{Q} = \alpha_Q \tilde{I}_1 + \tilde{J}_2^{1/2} - [k + k_\infty(1 - e^{-b\lambda})] \tag{1.25}$$

where α_Q is the dilatancy constant for the non-associated flow rule if $\alpha_Q \neq \alpha_F$.

The following form of plastic damage potential function (F) is adopted to have non-associate plastic flow in the effective stress space:

$$F = \tilde{Q} + \frac{s}{s+1} \left(\frac{Y}{S} \right)^{s+1} (1 - D)^\phi \tag{1.26}$$

where s, S, and ϕ are material parameters; D represents a damage variable; Y represents the damage conjugate force.

The experimental results of the stress-strain curves of concrete-like materials highly depend on the stress triaxiality. The phenomena of stress-triaxiality dependency of the stress-strain curves exists in engineering for a wide range of materials, such as geomaterials, ceramics, composites, and metals. To simulate this kind of phenomena, the stress triaxiality was used by several references (Alves and Jones, 1999; Borvik *et al.*, 2003; Horstemeyer *et al.*, 2000; Li *et al.*, 2002). The forms of stress triaxiality expressions differ from one to another: Alves and Jones (1999), Borvik *et al.* (2003), and Horstemeyer *et al.* (2000) define stress triaxiality explicitly, whereas Li *et al.* (2002) implicitly account for the stress triaxiality influence in its plastic hardening law by using the stress invariants I_1 and J_2.

For the convenience of model formulation and with a reference to the conventional expressions adopted in several references (Etse and Willam, 1999; Sfer *et al.*, 2002), the stress triaxiality R is defined here as:

$$R = |I_1/\sqrt{3}/\sqrt{2J_2}|, \quad J_2 \neq 0 \tag{1.27}$$

The stress triaxiality γ is introduced into the damage plastic loading condition and the damage plastic potential function in the following form:

$$\tilde{f} = \alpha_{QF} \tilde{I}_1 + \tilde{J}_2^{1/2} - [k + Rk_\infty(1 - e^{-bR\lambda})] = 0 \tag{1.28}$$

$$F = \tilde{Q} + \frac{SR}{s+1} \left(\frac{Y}{SR} \right)^{S+1} (1 - D)^\phi \tag{1.29}$$

where:

$$\tilde{Q} = \alpha_{QF} \tilde{I}_1 + \tilde{J}_2^{1/2} - [k + Rk_\infty(1 - e^{-bR\lambda})] \tag{1.30}$$

Consequently, plastic strain increment is obtained by:

$$\dot{\varepsilon}_{ij}^P = \dot{\lambda} \frac{\partial F}{\partial \sigma_{ij}} = \dot{\lambda} \frac{\partial \tilde{Q}}{\partial \sigma_{ij}} \tag{1.31}$$

With:

$$\frac{\partial \tilde{Q}}{\partial \sigma_{ij}} = \frac{1}{(1-D)} \left(\alpha_Q \delta_{ij} + \frac{S_{ij}}{2\sqrt{J_2}} \right) - k_\infty \left(\frac{\partial R}{\partial \sigma_{ij}} - \frac{\partial R}{\partial \sigma_{ij}} e^{-bR\lambda} + Rb\lambda e^{-bR\lambda} \frac{\partial R}{\partial \sigma_{ij}} \right) \tag{1.32}$$

With Equation (1.27), the following is derived:

$$\frac{\partial R}{\partial \sigma_{ij}} = (-1)^\eta \left[\frac{\delta_{ij}}{\sqrt{6J_2}} - \frac{I_1 S_{ij}}{2\sqrt{6}} (J_2)^{-3/2} \right], \quad \eta = \begin{cases} 1, & \text{if } I_1 < 0 \\ 2, & \text{if } I_1 \geq 0 \end{cases} \tag{1.33}$$

The damage evolution law can be derived as:

$$\dot{D} = \dot{\lambda}\left(\frac{Y}{SR}\right)^{S}(1-D)^{\phi} = \dot{\lambda}\bar{Y} \qquad (1.34)$$

The damage conjugate force Y in Equation (1.34) can be expressed as:

$$\bar{Y} = \left[\frac{(1-D)E^0_{ijkl}\varepsilon^e_{ij}\varepsilon^e_{kl}}{SR}\right]^s (1-D)^{\phi} \qquad (1.35)$$

The parameters used in this model are E, v, k, α_{F}, and α_{Q} for plasticity, and s, S, and ϕ for damage. Stress triaxiality (R) is a special variable introduced in this model.

1.4.2.3 *Constitutive behavior for a finite displacement increment* $\Delta\varepsilon_{ij}$

With the constitutive model previously described, the constitutive behavior can be derived for a known initial stress state (σ_{ij}, ε^p_{ij}, and D) and a given strain increment ($\Delta\varepsilon_{ij}$). The stress increment can be obtained by making the total differential operation over the total stress tensor in Equation (1.23), and a subsequent linearization over the time increment (Δt), thus:

$$\Delta\sigma_{ij} = E^0_{ijkl}(1-D)^2(\Delta\varepsilon_{kl} - \Delta\varepsilon^p_{kl}) - 2E^0_{ijkl}(1-D)(\varepsilon_{kl} - \varepsilon^p_{kl})\Delta D$$

$$= -\Delta\lambda\left[E^0_{ijkl}(1-D)^2\frac{\partial\tilde{Q}}{\partial\sigma_{kl}} + 2E^0_{ijkl}(1-D)\varepsilon^e_{kl}\bar{Y}\right] + E^0_{ijkl}(1-D)^2\Delta\varepsilon_{kl} \qquad (1.36)$$

where superscript 0 indicates the value of parameter at the initial state of calculation. It also means intact material at very beginning of the calculation.

With these equations, the following equation is obtained:

$$\sigma_{ij} = \sigma^0_{ij} + \Delta\sigma_{ij} = E^0_{ijkl}(1-D^0)^2(\varepsilon^0_{kl} - \Delta\varepsilon^{p^0}_{kl}) + E^0_{ijkl}(1-D^0)^2\Delta\varepsilon_{kl}$$

$$-\Delta\lambda\left[E^0_{ijkl}(1-D^0)^2\frac{\partial\tilde{Q}}{\partial\sigma^0_{kl}} + 2E^0_{ijkl}(1-D^0)(\varepsilon^0_{kl} - \varepsilon^{p^0}_{kl})\bar{Y}\right]^0 \qquad (1.37)$$

The plastic damage multiplier ($\dot{\lambda}$) can be determined explicitly with the following consistency condition:

$$\tilde{f} = \dot{\tilde{f}} = 0 \qquad (1.38)$$

Consequently, the total differential of f is obtained with the following equation:

$$\tilde{df} = \frac{\partial\tilde{f}}{\partial\sigma_{ij}}d\sigma_{ij} + \frac{\partial\tilde{f}}{\partial D}dD + \frac{\partial\tilde{f}}{\partial\lambda}d\lambda = \frac{\partial\tilde{f}}{\partial\sigma_{ij}}(1-D)^2E^0_{ijkl}d\varepsilon_{kl}$$

$$-d\lambda\left[\frac{\partial\tilde{f}}{\partial\sigma_{ij}}E^0_{ijkl}(1-D)^2\frac{\partial\tilde{Q}}{\partial\sigma_{kl}} + 2\frac{\partial\tilde{f}}{\partial\sigma_{ij}}E^0_{ijkl}(1-D)\varepsilon^e_{kl}\bar{Y} - \frac{\partial\tilde{f}}{\partial D}\bar{Y} - \frac{\partial\tilde{f}}{\partial\lambda}\right] \qquad (1.39)$$

Thus, the following expression for plastic multiplier can be derived:

$$d\lambda = \frac{\dfrac{\partial\tilde{f}}{\partial\sigma_{ij}}E^0_{ijkl}(1-D)^2}{\left[\dfrac{\partial\tilde{f}}{\partial\sigma_{ij}}E^0_{ijkl}(1-D)^2\dfrac{\partial\tilde{Q}}{\partial\sigma_{kl}} + 2\dfrac{\partial\tilde{f}}{\partial\sigma_{ij}}E^0_{ijkl}(1-D)\varepsilon^e_{kl}\bar{Y} - \dfrac{\partial\tilde{f}}{\partial D}\bar{Y} - \dfrac{\partial\tilde{f}}{\partial\lambda}\right]}d\varepsilon_{kl} \qquad (1.40)$$

For the sake of brevity, Equation (1.40) can be re-written as:

$$d\lambda = \frac{A_{kl}}{B}d\varepsilon_{kl} \qquad (1.41)$$

where:

$$
\begin{cases}
A_{kl} = \dfrac{\partial \tilde{f}}{\partial \sigma_{ij}} E^0_{ijkl}(1-D)^2 \\[3mm]
B = \dfrac{\partial \tilde{f}}{\partial \sigma_{ij}} E^0_{ijkl}(1-D)^2 \dfrac{\partial \tilde{Q}}{\partial \sigma_{kl}} + 2\dfrac{\partial \tilde{f}}{\partial \sigma_{ij}} E^0_{ijkl}(1-D)\varepsilon^e_{kl}\bar{Y} - \dfrac{\partial \tilde{f}}{\partial D}\bar{Y} - \dfrac{\partial \tilde{f}}{\partial \lambda}
\end{cases}
\tag{1.42}
$$

For a given strain increment ($\Delta\varepsilon_{ij}$), the stress tensor increment can be obtained by substituting Equation (1.42) into Equation (1.36) such that:

$$
\Delta\sigma_{ij} = \frac{A_{kl}}{B}\left[E^0_{ijrs}(1-D)^2 \frac{\partial \tilde{Q}}{\partial \sigma_{rs}} + 2E^0_{ijrs}(1-D)\varepsilon^e_{rs}\bar{Y} \right]\Delta\varepsilon_{kl} + E^0_{ijkl}(1-D)^2\Delta\varepsilon_{kl}
\tag{1.43}
$$

Therefore, the algorithmic tangential stiffness tensor can be deduced as:

$$
E^{epd}_{ijkl} = \frac{\partial \Delta\sigma_{ij}}{\partial \Delta\varepsilon_{kl}} = E^0_{ijkl}(1-D)^2 - \left[E^0_{ijrs}(1-D)^2 \frac{\partial \tilde{Q}}{\partial \sigma_{rs}} + 2E^0_{ijrs}(1-D)\varepsilon^e_{rs}\bar{Y} \right]\frac{A_{kl}}{B}
\tag{1.44}
$$

where superscript epd indicates elastoplastic damage tangential stiffness.

The elasto-plastic damage loading condition for a given strain increment ($\Delta\varepsilon_{ij}$) can be expressed conceptually in the effective stress space as:

$$
\tilde{f} = \tilde{f}^0 + \frac{\partial \tilde{f}}{\partial (\Delta\lambda)}, \quad \Delta\lambda \le 0
\tag{1.45}
$$

where $\tilde{f}^0$ is the value of yielding function at the starting effective stress state ($\tilde{\sigma}^0_{ij}$). With Equation (1.28), the following relationship is obtained:

$$
\frac{\partial \tilde{f}}{\partial (\Delta\lambda)} = \frac{\partial \tilde{f}}{\partial \sigma_{ij}} \frac{\partial \sigma_{ij}}{\partial (\Delta\lambda)} + \frac{(\alpha I_1 + \sqrt{J_2})}{(1-D)^2} \frac{\partial D}{\partial (\Delta\lambda)} + \frac{\partial \tilde{f}}{\partial \lambda} \frac{\partial \lambda}{\partial (\Delta\lambda)}
\tag{1.46}
$$

With Equation (1.23) and Equation (1.24), the tensors and vectors on the right side of Equation (1.46) are obtained by:

$$
\frac{\partial \tilde{f}}{\partial \sigma_{ij}} = \frac{1}{(1-D)}\left(\alpha\delta_{ij} + \frac{S_{ij}}{2\sqrt{J_2}} \right) - k_\infty \left(\frac{\partial R}{\partial \sigma_{ij}} - \frac{\partial R}{\partial \sigma_{ij}} e^{-bR\lambda} + Rb\lambda e^{-bR\lambda} \frac{\partial R}{\partial \sigma_{ij}} \right)
\tag{1.47}
$$

$$
\frac{\partial \sigma_{ij}}{\partial (\Delta\lambda)} = -\left[E^0_{ijkl}(1-D)^2 \frac{\partial \tilde{Q}}{\partial \sigma_{kl}} + 2E^0_{ijkl}(1-D)\varepsilon^e_{kl}\left(\frac{Y}{SR} \right)^s (1-D)^\phi \right]
\tag{1.48}
$$

$$
Y = (1-D)E^0_{ijkl}\varepsilon^e_{ij}\varepsilon^e_{kl}
\tag{1.49}
$$

$$
\frac{\partial \tilde{f}}{\partial (\lambda)} = -bR^2 k_\infty e^{-bR\lambda}
\tag{1.50}
$$

$$
\frac{\partial D}{\partial (\Delta\lambda)} = \left(\frac{Y}{SR} \right)^s (1-D)^\phi
\tag{1.51}
$$

The formulation of the Newton-Raphson iteration equation between $\Delta\lambda$ and $\tilde{f}$ is formed as:

$$
\Delta\lambda = \Delta\lambda_0 - \tilde{f}\left(\frac{\partial \tilde{f}}{\partial \lambda} \right)^{-1}
\tag{1.52}
$$

where $\tilde{f}^0$ is the value of the yielding function at the beginning effective stress state ($\tilde{\sigma}^0_{ij}$).

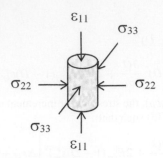

Figure 1.5   Illustration of the mixed loading condition.

1.4.3   Numerical validation of constitutive model at the local level

In this subsection, a driver subroutine is designed to validate a 3D constitutive model at the local level, i.e., for a material point only, with reference to the algorithm proposed in Hashash *et al.* (2002). Figure 1.5 illustrates its principle as a mixed loading condition is applied with $\varepsilon_{11} = \varepsilon_{11}(t)$ and $\sigma_{22} = \sigma_{33} =$ constant, which means that a strain loading will be applied incrementally under a constant stress confinement in the other two directions. The strain loading is applied elastically in direction 11, whereas the self-equilibrium mechanism at this material point will result in a variation of lateral strains nonlinearly (for the sake of damage) to maintain a constant lateral confinement. The details of the numerical calculations will be given in the following content.

1.4.3.1   *Iteration procedure for the constitutive validation: external equilibrium iteration*
The function of the external equilibrium iteration is for a known stress and strain state and a set of internal variables (σ_{ij}, ε_{ij}, ε_{ij}^{p}, and D), applying a load increment ($\Delta\varepsilon_1$, $\Delta\sigma_2$, and $\Delta\sigma_3$) to quasi-elastically determine the response of $\Delta\sigma_1$, $\Delta\varepsilon_2$, $\Delta\varepsilon_3$, and D by an iterative procedure.

The following quasi-elastic equations (i.e., elastic relationship for a finite time increment Δt) are adopted in the calculation:

$$\begin{Bmatrix} \Delta\varepsilon_{22} \\ \Delta\varepsilon_{33} \end{Bmatrix} = \begin{bmatrix} E_{2222} & E_{2233} \\ E_{3322} & E_{3333} \end{bmatrix} \left[\begin{Bmatrix} \Delta\sigma_{22} \\ \Delta\sigma_{33} \end{Bmatrix} - \begin{Bmatrix} E_{2211} \\ E_{3311} \end{Bmatrix} \Delta\varepsilon_{11} \right] \tag{1.53}$$

where E_{ijkl} are components of the elasticity tensor of damaged material, which is expressed in Equation (1.44). Figure 1.6 illustrates the principle of the external equilibrium iteration.

In Figure 1.6, the UMAT is the constitutive module that checks the elasto-plastic loading state and performs elastic and/or elasto-plastic damage calculations. The subroutine CONSTITUERE will be used in the UMAT. The function of subroutine CONSTITUERE is to perform the constitutive integration and will be introduced in detail in the following subsection. The elasto-plastic-damage loading stiffness (E_{ijkl}^{epd}), also known as algorithmic tangential stiffness, will be updated after every iteration, and will be used in the quasi-elastic calculation of $\Delta\varepsilon_2$ and $\Delta\varepsilon_3$ at every first-iteration step at each of the loading increments. Superscript ep of variables in Figure 1.6 means values of stress increment calculated with elasto-plastic damage constitutive relationship. Superscript r of variables shown in Figure 1.6 means the residual stress calculated in this increment. Superscript k is the current iteration number.

1.4.3.2   *Iteration procedure for the constitutive validation: internal elasto-plastic damage iteration*
By using the "fixed-point" method described in Chaboche and Cailletaud (1996), for a given finite strain ($\Delta\varepsilon_{ij}$), the only unknown in the elasto-plastic damage calculation at a local level is the plastic-damage multiplier ($\Delta\lambda$).

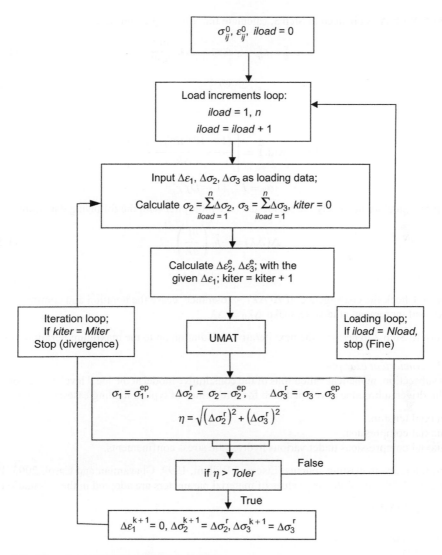

Figure 1.6   Flow chart of global equilibrium iteration.

The solution steps adopted in the procedure of CONSTITUERE subroutine include the following:

- Step 1: Initiate the stress state and the state of all internal variables ($\sigma_{ij}^0, \varepsilon_{ij}^0, \varepsilon_{ij}^p$, and D_0). Both superscript 0 and subscript 0 means values of variables at initial state. And it is true for the following subscript and superscript 0 used in this section.
- Step 2: Apply the strain increment $\Delta\varepsilon_{ij}$, with $\Delta\varepsilon_{ij} = 0$ for $i \neq j$ obtained from the outer global equilibrium iteration.
- Step 3: Calculate $\Delta\lambda_0$ with the given initial stress state, strain increment, and linearized Equation (1.40).

- Step 4: With $\Delta\lambda_0$ obtained in Step 3, calculate the following quantities:

$$\varepsilon_{ij}^{e} = \varepsilon_{ij}^{e(0)} + \Delta\varepsilon_{ij} - \Delta\lambda_0 \frac{\partial \tilde{Q}}{\partial \sigma_{ij}^0} \tag{1.54}$$

$$\varepsilon_{ij} = \varepsilon_{ij}^{(0)} + \Delta\varepsilon_{ij} \tag{1.55}$$

$$\varepsilon_{ij}^{p} = \varepsilon_{ij} - \varepsilon_{ij}^{e} \tag{1.56}$$

$$D = D^{(0)} + \Delta\lambda_0 \bar{Y}, \text{ with } \bar{Y} = \left[\frac{(1 - D^{(0)}) E_{ijkl}^0 \varepsilon_{ij}^e \varepsilon_{kl}^e}{SR} \right]^2 (1 - D^{(0)})^\phi \tag{1.57}$$

$$\sigma_{ij} = E_{ijkl}^0 (1 - D)^2 \varepsilon_{ij}^e \tag{1.58}$$

- Step 5: Calculate iteratively the plastic-damage multiplier with the following equations:

$$\Delta(\Delta\lambda) = -\tilde{f}_0 \left(\frac{\partial \tilde{f}}{\partial \lambda} \right)^{-1} \tag{1.59}$$

$$\Delta\lambda = \Delta\lambda_0 + \Delta(\Delta\lambda) \tag{1.60}$$

- Step 6: Check the convergence: if $\Delta(\Delta\lambda) \leq$ tolerance, cease the iteration and continue to the next load increment; otherwise, make $\Delta\lambda_0 = \Delta\lambda$

Return to Step 2 to perform the next iterative calculation up to the maximum iteration limit.

1.4.3.3   *Numerical examples*

In this subsection, numerical validations of the constitutive model at the local level are performed with the driver subroutine developed here for three kinds of typical loading cases:

- Uniaxial tension.
- Uniaxial compression.
- Uniaxial compressions under various hydrostatic stress confinements.

With reference to existing sources (Etse and Willam, 1999; Ghavamian and Carol, 2003; Lee and Fenves, 1998), the following values of material parameters are adopted in the calculation:

- $E = 31{,}140$ MPa.
- $v = 0.2$.
- $\alpha_F = \alpha_Q = 0.15$.
- $K = 2.0$ MPa.
- $s = 1$.
- $S = 10^{-16}$ for tension and 4×10^{-5} for compression.
- $\phi = -1.0$.
- $b = 88$ for tension and 500 for compression.
- $k_\infty = 100$ MPa for tension and 248 MPa for compression.
- Tolerance $= 10^{-20}$ for internal iteration (i.e., for $\Delta(\Delta\lambda)$) and 10^{-4} for external iteration (i.e., for constant lateral stress confinement).

Figure 1.7 shows the stress-stain behavior under uniaxial tension of the model. No plastic hardening behaviors are observed in this case. A comparison between the experimental data (Gopalaratnam and Shah, 1985) and the numerical results of the stress-strain response under uniaxial tension indicates that the pre-peak stress-strain behavior can be reliably predicated, whereas the post-peak behavior can only be predicted with a reasonable accuracy by the proposed model. Figure 1.8 shows the numerical results of the response of the lateral strain and volumetric strain. The damage response in Figure 1.9 shows that damage value asymptotically tends to be a maximum of 1.0 with the increase of strain loading.

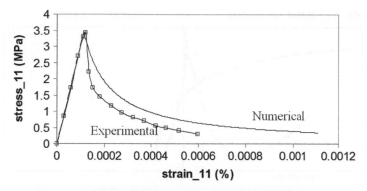

Figure 1.7   Stress-strain behavior under uniaxial tension: comparison between numerical and experimental results.

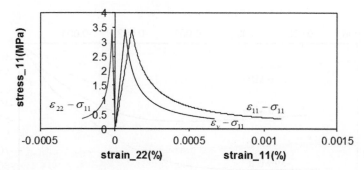

Figure 1.8   Stress-strain behaviors under uniaxial tension: lateral and volumetric responses.

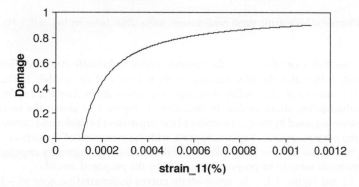

Figure 1.9   Damage evolution under uniaxial tension.

The stress-stain behavior and damage evolution response of the proposed model under uniaxial compression are shown in the following Figures 1.10 to 1.16. Comparison between the experimental data given by Karsan and Jirsa (1969) and the numerical results of the stress-strain response under uniaxial compression in Figure 1.10 indicates that the pre-peak behavior can be reliably predicated, and the post-peak behavior can be predicted with a reasonable degree of accuracy. The numerical results of the response of the lateral strain and volumetric strain are shown in Figure 1.10 and indicate the dilatancy property of the model; there is a saturation of dilatancy at which the volumetric strain ceases to increase. Figure 1.14 presents the corresponding damage evolution behavior.

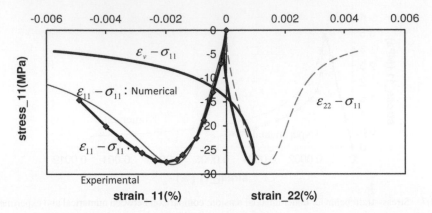

Figure 1.10   Stress-strain behaviors under uniaxial compression.

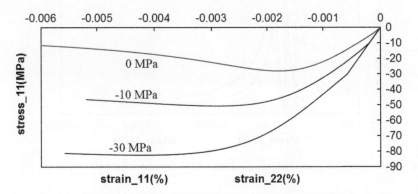

Figure 1.11   Influence of hydrostatic stress confinement: stress-strain behavior ($\sigma_m = 0, -10, -30$ MPa).

The stress-strain behavior of a model for concrete under hydrostatic stress confinement is an important aspect; it indicates the pressure-sensitivity behavior of the model. In the numerical tests described in this section, the loading procedure is given as the hydrostatic confinement (i.e., $\sigma_m \mathbf{I}$) and is applied before strain loading in direction 11. Figure 1.11 shows the variation of the stress-strain response caused by the confinement of the stress-stain behavior and damage evolution response with the other parameters kept unchanged; with the increment of the stress confinement, the softening phenomena become progressively weaker. The stress-triaxiality dependent plastic-hardening phenomena seem to be properly simulated by the proposed model.

In Figure 1.12 and Figure 1.13, the stress-strain curves under confinement of -10 MPa and -30 MPa are given, respectively, with the responses of $\varepsilon_{22} - \sigma_{11}$ and $\varepsilon_v - \sigma_{11}$. The dilatancy phenomena become weaker with the increase of confinement pressure. The pre-peak nonlinearity of the stress-strain curve seems reasonably simulated.

To illustrate the proposed model in more depth (Figs. 1.15 and 1.16), the peak strength envelopes obtained numerically with the proposed model are provided. Because of the different parameter values adopted in the calculation for tension and compression, Figure 1.15 shows that the shape of the tensile-strength envelope is quite different from the shape of the compressive-strength envelope. The axial points in Figure 1.15 are obtained analytically by using the Drucker-Prager condition directly, because there is no pre-peak nonlinearity for the tensile case, which is judged by criterion $I_1 \geq 0$. In Figure 1.16, the envelope of compression peak strengths under very high confinement up to -200 MPa is shown to demonstrate the validity of the model for a wide range of hydrostatic-stress confinement.

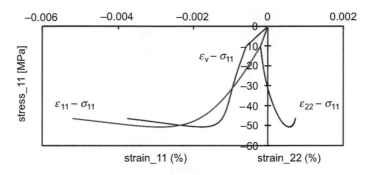

Figure 1.12   Stress-strain behaviors under compression with $-10\,$MPa confinement.

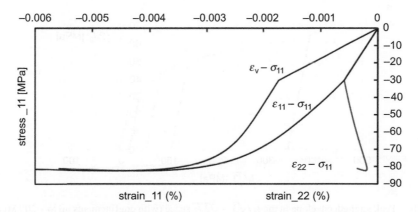

Figure 1.13   Stress-strain behaviors under compression with $-30\,$MPa confinement.

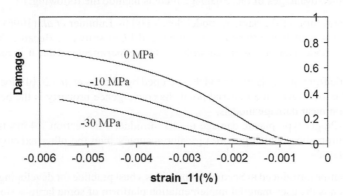

Figure 1.14   Damage-strain behaviors under compression with various stress confinements.

1.4.4   Concluding remarks

This section introduced several typical CDM models. The damage variables of these models are in scalar form. Because of the high nonlinearity that exists among the model parameters and the experimental phenomena, it seems necessary to select the values of constitutive parameters by some kind of technique of inverse analysis.

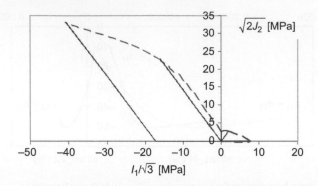

Figure 1.15   Peak strength envelope in the $I_1/\sqrt{3} - \sqrt{2J_2}$ space (with confinements up to -30 MPa).

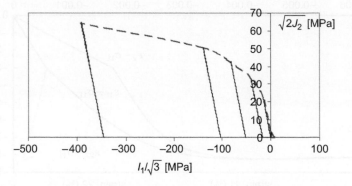

Figure 1.16   Peak strength envelope in the $I_1/\sqrt{3} - \sqrt{2J_2}$ space (with confinements up to -200 MPa).

In summary, the advantages of these damage models include the following:

- The incremental form of the damage model developed by Lubliner *et al.* (1996) and Lee and Fenves (1998) is available in the current commercial FEM software. Because the parameter values of this model can be calibrated with existing experimental data, it is relatively easy to use.
- The holonomic form of the damage model developed by Mazars *et al.* (2014) does not require iteration at the local level of a material point. Its convergence property is the best among all the current nonlinear damage models.
- The Drucker-Prager type plastic damage model introduced in Section 1.4 has triaxial-stress dependency properties, which provides greater accuracy than the other existing models for rock-like materials.

The driver subroutine introduced in Section 1.4 provides a best practice for developing and validating UMAT, which is the user-material implementation platform of some large-scale commercial FEM software.

CHAPTER 2

Optimizing multistage hydraulic-fracturing design based on 3D continuum damage mechanics analysis

This chapter uses a real three-dimensional (3D) method to optimize the design of stage intervals in multistage hydraulic fracturing and the design of horizontal well spacing in unconventional oil and gas resources.

This method uses continuum damage mechanics as its theoretic basis for modeling fractures and their propagation under stimulation injection. The volumetric density of cracks created by injection fluid along with those natural fractures can be represented by a set of two scalar-damage variables.

This method can provide a solution for fracture propagation in two horizontal directions: the axial and lateral directions of the trajectory. This set of fracture propagation numerical solutions is used to optimize the design of stage intervals for multistage hydraulic fracturing; it is also used to optimize the design for well spacing in parallel horizontal wells, which is known as a zipper fracture. The determination of optimized stage intervals considers the overlapping effect of two nearby stimulation stages.

This method is known as "real 3D" because the distribution of the continuum damage variable, which represents clouds of fractures, is 3D volumetric in all three directions of the model.

2.1 INTRODUCTION

This chapter demonstrates how to build a real 3D method to optimize the design of stage intervals for multistage hydraulic fracturing and to optimize the design of well spacing for horizontal wells for unconventional oil and gas resources. The term "real 3D" means the fracture propagation occurs in vertical and transverse directions, and that the fracture propagation in the lateral direction will be calculated. This process makes the analysis of the fracture propagation a full 3D analysis. Continuum damage mechanics (CDM) is a theoretical branch of solid mechanics that investigates the development of crack clouds, rather than the development of a single crack (Lemaitre and Chaboche, 1994).

As a major method of reservoir stimulation, hydraulic fracturing has been investigated since the 1950s (Cleary, 1980; Khristianovic and Zheltov, 1955; Warpinski, 1985). Because of the rapid development of unconventional petroleum resources in recent years, the investigation of hydraulic fracturing has again attracted the interest of many researchers (Bagherian *et al.*, 2010; Bahrami and Mortazavi, 2008; Cipolla *et al.*, 2010; Ehlig-Economides *et al.*, 2006; Mayerhofer *et al.*, 2010; Shaoul *et al.*, 2011; Soliman *et al.*, 2004; Warpinski, 2013).

Two critical parameters in the hydraulic fracturing design of unconventional petroleum resources are the stage interval of the multistage stimulation and the well spacing between nearby horizontal wells. The proper selection of values for these two parameters is a key component to obtain successful oil and gas production from unconventional resources, such as tight-sand oil and shale gas. The method used in the current design of the stage intervals of multistage hydraulic fracturing is based on the fracture propagation analysis of a pseudo 3D fracture analysis. The software used in the current fracture analysis tools, such as StimPlan™ (Smith, 2010), uses a 3D planar fracture model. The outputs of the fracturing simulation are primarily the length and height of the 3D planar fracture. Little information about the lateral direction can be obtained with this type of analytical tool. The stress shadow method (Nagel *et al.*, 2013) investigates the stress distribution around a given crack under a given fluid pressure at crack surfaces; it was used by a few researchers (Nagel *et al.*, 2013; Skomorowski *et al.*, 2015) for the design of stage intervals of multistage hydraulic fracturing and the optimized design of well spacing for horizontal wells of unconventional oil and gas resources. However, in its analysis process, the stress shadow method does not calculate fracture propagation under hydraulic fracturing. Consequently, it provides no information about fracture propagation and represents a very rough approximation method for determining the stage interval.

The method proposed in this chapter uses CDM (Lee and Fenves, 1998; Swoboda *et al.*, 1998) as its theoretical basis for fracture modeling and fracture propagation under stimulation injection. In rock mechanics, the mechanical damage variable (DV) is interpreted as an index of material continuity, which varies from 0 for intact rock to 1 for completely separated, broken rock. The volumetric density of cracks created by injection fluid can be represented by a set of two-scalar DVs. This method can thus provide a fracture propagation solution in three directions of a 3D volume. The solution of fracture propagation under injection in the two horizontal directions will be used primarily for the optimized design for stage intervals of multistage hydraulic fracturing, as well as the optimized design of well spacing for horizontal wells.

Data used in the numerical application in this chapter is for the purpose of illustrating the procedure only.

2.2 THE WORKFLOW

Figure 2.1 illustrates the workflow and steps of this 3D method for the optimized design of stage intervals of multistage hydraulic fracturing. The procedure details include the following:

- Step 1: Build a 3D global model for the field and calculate the initial geostress distribution with a 3D finite element tool, such as Abaqus. The scale of the field model is usually in kilometers.
- Step 2: Build a reservoir level submodel at the scale of hundreds of meters. The center of the submodel should be located at the position of the perforation section. The values of stress components and displacement vectors at the locations of the submodel external surfaces should be subtracted from the 3D numerical results of the field model obtained in Step 1 and applied as boundary conditions to the submodel.
- Step 3: Perform a poro-elastoplastic damage calculation with the finite element method to determine the solution for the distribution of the continuum DVs under stimulation injection loads.
- Step 4: Determine the effective fracture length with the critical value of the synthetic DV and further determine the optimized stage interval and well spacing, respectively, with reference to the numerical results of the continuum damage distribution generated by the stimulation. The synthetic DV combines both tensile DV and compressive/crushing DV.

The effective fracture is the fracture with an opening that is large enough to hold proppant; consequently, it can contribute to oil and gas production. Usually, the larger the opening of a propped fracture, the greater its production capacity will be. In general, from the injection point to the front tip of a fracture generated by stimulation injection, the opening of the fracture decreases from a maximum value to zero. The critical value of the synthetic DV is the DV value that can represent the opening of an effective fracture. If the value of the synthetic DV reaches this critical value, this point will be regarded as a part of an effective fracture. The half-length of an effective fracture is the distance between the injection point and the farthest point of the effective fracture. This point is illustrated with an example provided in the following sections.

Microseismic data obtained from the stimulation of offset wells of the same field is used as reference to determine the critical value of the DV of a given target formation. This procedure is performed by making a numerical calculation of the hydraulic fracturing with the offset well

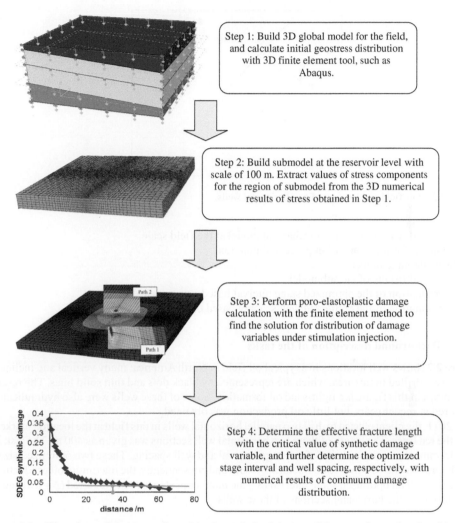

Step 1: Build 3D global model for the field, and calculate initial geostress distribution with 3D finite element tool, such as Abaqus.

Step 2: Build submodel at the reservoir level with scale of 100 m. Extract values of stress components for the region of submodel from the 3D numerical results of stress obtained in Step 1.

Step 3: Perform poro-elastoplastic damage calculation with the finite element method to find the solution for distribution of damage variables under stimulation injection.

Step 4: Determine the effective fracture length with the critical value of synthetic damage variable, and further determine the optimized stage interval and well spacing, respectively, with numerical results of continuum damage distribution.

Figure 2.1   Flow chart of the steps adopted in the optimized design of the stage intervals of multistage hydraulic fracturing.

information, and then comparing the effective fracture length with the interpreted fracture on the basis of microseismic data.

2.3   VALIDATION EXAMPLE

The following subsections present an example of the application of the 3D method proposed for the optimized design of a stage interval of multistage hydraulic fracturing and for the optimized design of well spacing for horizontal wells.

The proposed workflow shown in Figure 2.1 is used to determine the optimized stage interval and well spacing. In the following sections, details about the related calculations and analysis are provided. The sequence of these sections includes the following:

- Background description of the tasks.
- 3D geomechanical model at the field scale.

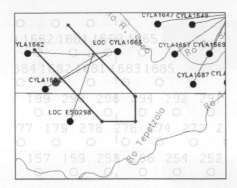

Figure 2.2   Oil field and location of two horizontal wells.

- Numerical results of the geomechanical model at the field scale.
- Submodel for the stimulation process simulation.
- Plastic damage model.
- Numerical results of the submodel.
- Determination of the optimized stage interval based on numerical solutions.
- Determination of the optimized well spacing based on numerical solutions.

2.3.1   Background description of the tasks

Figure 2.2 shows well locations in a typical oil field in North America; many vertical and inclined wells were drilled in this area, which are represented by black dots and thin solid lines. The reservoir shown in this figure is a tight-sand oil formation. Some of these wells were also hydraulically fractured in recent years, but little oil production was obtained.

In 2011, the client elected to drill two parallel horizontal wells in this field at the location marked with the red lines. The length of these two horizontal well sections was given as 400 m. The task is to determine the optimized size of the stage interval and well spacing. These two optimized sizes should provide the best hydraulic fracturing result, and consequently, the maximum oil production quantity. With an optimized stage interval (S), the number of stages to be fractured (N) will equal $N = 400/S$ for the horizontal sections of these wells.

2.3.2   3D geomechanical model at field scale

A 3D finite element model was built to obtain the geostress distribution of this field. A numerical stress analysis was performed at the field scale. Figure 2.3 shows the model geometry. The following factors of the model were adopted in the model definition:

- Y-direction of the model was selected as the direction of the maximum principal stress, which has an angle of $+14°$ to the north, as shown in Figure 2.3.
- The model scale is 10×10 km (width and length) at a thickness/depth of 6 km.
- The top of the wellbore is in the center of the model.
- The inclination angle of the formation sediment layer is set as $6.5°$.
- 5000 cubic elements with coupled displacement and pore pressure (C3D20RP) were used in discretization of the mesh.

Figure 2.4 shows the boundary conditions; the key characteristics include the following:

- The normal displacement constraints are applied to all lateral surfaces and to the bottom surface; the top surface is set as a free surface.
- A simplification has been adopted in that the variation of the top surface elevation is neglected.

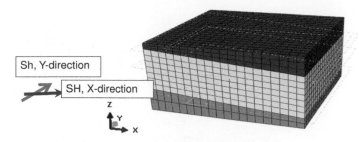

Figure 2.3   3D model geometry and directions.

Figure 2.4   3D model showing boundary conditions.

- Initial conditions include the following:
  - The initial geostress field is applied with an effective stress ratio of $k0 = 0.6$ and a tectonic factor of $tf = 0.5$. The initial value will be balanced with the formation gravity load and automatically modified by the equilibrium equations.
  - An initial pore pressure of 20.7 MPa (3000 psi) is applied to the reservoir formations.
- Loads:
  - Gravity load is applied to the model; it will be used to balance the initial geostress field with the existence of pore pressure.
  - Injection load is not applied in this process; it is applied only to the submodel.

Values of material parameters used in the global model include the following:

- Young's modulus $E = 10$ GPa; the value used in this model is for an un-drained material status.
- Poisson's ratio $\nu = 0.23$; the value used in this model is for an un-drained material status.
- Permeability $k = 1$ mD; hydraulic conductivity $\hat{K} = 9.8 \times 10^{-9}$ m/s.
- Density $\rho = 2300$ kg/m^3

2.3.3   Numerical results of the geomechanical model at field scale

As a visualization of the principal stress directions, Figure 2.5a-c show XoY-plane views of the mechanical variables. Figure 2.5a shows the maximum compressive stress at $TVD = 1700$ m, Figure 2.5b shows the minimum compressive stress, and Figure 2.5c shows the medium compressive stress.

2.3.4   Submodel for stimulation process simulation

Submodeling techniques are used to accommodate the field-to-borehole-section scale discrepancy. The concept of the submodeling technique includes using a large-scale global model to produce boundary conditions for a smaller scale submodel. In this way, the hierarchical levels of the submodel are not limited. Using this approach, a highly inclusive field-scale analysis can

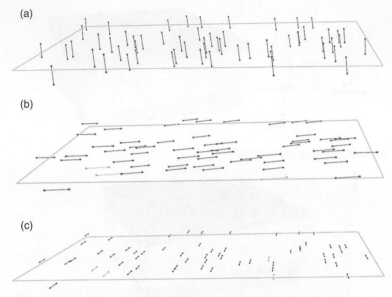

Figure 2.5　Numerical results of 3D model: visualization of the principal directions of the initial stress tensor.

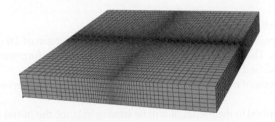

Figure 2.6　Mesh and geometry of the submodel.

be linked to a very detailed stress analysis at a much smaller borehole scale. The benefits are bidirectional, with both the larger and smaller scale simulations benefiting from the linkage.

The geometry of the submodel is set with a width and length of 200 m and a thickness of 20 m. The calculation is focused on the lateral scope of the fractured volume generated by fluid injection. Therefore, in the vertical direction, the analysis of the mechanical behavior of the model was simplified by taking only a thin slice shape in 3D space.

A total of 21,600 3D eight-node continuum elements (C3D8RPH) are used in the discretization of the model. Figure 2.6 shows the mesh of the submodel.

The task of this calculation is to estimate the injection effect with the previously mentioned parameter values. This calculation will simulate the fracture generation under a bottomhole pressure (*Btmh*) of 39.5 MPa (5730 psi) after an initial peak value, with a perforation section length of 10 m, along with the following conditions:

- The initial pore pressure of the reservoir is 20.7 MPa (3000 psi).
- The boundary conditions of displacement and pore pressure are taken from the numerical results of the global model.
- The initial geostress values are taken from the numerical results of the global model, corresponding to $TVD = 1700$ m.

2.3.5   **The plastic damage model**

The damage model used to model the fracture development in the submodel is a plasticity-based scalar continuum damage model. Chapter 1 includes details of the plastic-damage model theoretical description.

Table 2.1 lists the values of material parameters used in this calculation.

Figure 2.8 shows the pore pressure distribution during the stable injection process within a central sectional view of the XoZ plane. This figure also shows the distribution of the tensile *DV* within a central, sectional view of the XoY plane.

Two paths were selected to visualize continuum damage distribution values. Path 1 was selected, as shown in Figure 2.9, in the *X*-direction, and Path 2 is in the *Y*-direction. The points on these paths are nodes of the finite element mesh. The values of the continuum *DV*s occurred at the points (nodes) along Path 1 and Path 2 shown in Figure 2.10 and Figure 2.11, respectively.

Table 2.1   Material parameters of plastic damage model.

Material name = tight sand		
Young's modulus [Pa]	Poisson's ratio	Density [kg/m^3]
1.00×10^{10}	0.23	2380

Concrete damaged plasticity				
Dilatancy ψ	Eccentricity	σ_{b0}/σ_{c0} ratio	kc	Viscosity μ [Pa s]
28°	0.1	1.16	0.7	1.2

Concrete compression hardening	
Yield stress [Pa]	Inelastic strain
2.40×10^6	0
2.92×10^6	0.00004
3.17×10^6	0.00008
3.24×10^6	0.00012
3.18×10^6	0.00016
3.04×10^6	0.0002
2.85×10^6	0.00024
2.19×10^6	0.00036
1.49×10^6	0.0005
2.95×10^5	0.001
2.80×10^5	0.002

Concrete tension stiffening	
Tensile strength [Pa]	Cracking strain
3.70×10^3	0
3.00×10^3	5.00×10^{-8}
2.30×10^3	5.00×10^{-7}
1.30×10^3	5.00×10^{-6}
330	5.00×10^{-5}
133	5.00×10^{-4}

(Continued)

Table 2.1   (Continued)

Concrete compression damage

Compressive damage	Inelastic strain
0	0
0.1299	0.00004
0.2429	0.00008
0.3412	0.00012
0.4267	0.00016
0.5012	0.0002
0.566	0.00024
0.714	0.00036
0.8243	0.0005
0.9691	0.001
0.99	0.002

Concrete tension damage

Tensile damage	Cracking strain
0	0
0.1	5.00×10^{-8}
0.15	5.00×10^{-7}
0.2	5.00×10^{-6}
0.25	5.00×10^{-5}
0.9	5.00×10^{-4}

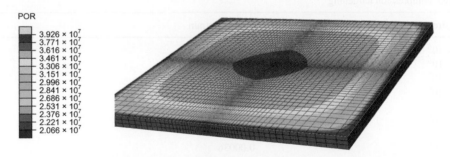

Figure 2.7   Pore pressure distribution after stimulation of the submodel [Pa].

Figure 2.8   Pore pressure distribution during stimulation: XoZ plane, sectional view.

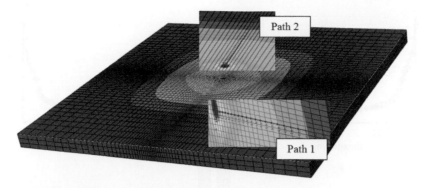

Figure 2.9   Illustration of Path 1 and Path 2 in XoY sectional view.

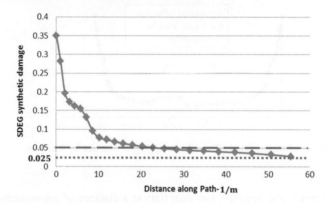

Figure 2.10   Distribution of synthetic damage along Path 1.

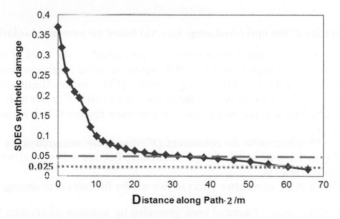

Figure 2.11   Distribution of synthetic damage along Path 2.

Figure 2.10 shows that the maximum continuum damage value on the nodes along Path 1 is 0.35 and occurs at the point of injection. The continuum damage value decreases with distance away from the injection point, and becomes less than 0.05 at a distance of approximately 24 m; it is less than 0.025 at a distance of 55 m.

Figure 2.11 shows that the maximum continuum damage value on nodes along Path 2 is 0.375. It occurs at the point of injection. The continuum damage value decreases with distance away

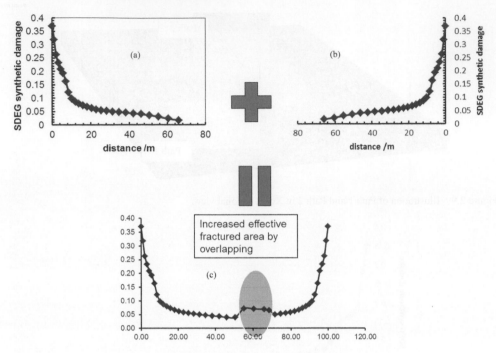

Figure 2.12   Illustration of accumulated effect of overlapping of two nearby fractures.

from the injection point, and becomes less than 0.05 at a distance of approximately 32 m; it is less than 0.025 at a distance of 63 m.

2.3.6   Determination of the optimized stage interval based on numerical solutions

The determination of the optimized stage interval is performed on the basis of the numerical solution shown in Figure 2.10 and Figure 2.11. With reference to the microseismic data obtained from offset wells, the critical value of the damage variable (*CDV*) for an effective fracture in this field is assumed to be 0.05. This means that the discontinuity index of the formation material is 5%. The point at which the continuum *DV* value at a location reaches 0.05 is regarded as a part of an effective fracture.

Consequently, with reference to the continuum *DV* values, *DV* occurred at the points along Path 1 shown in Figure 2.10; the half-length of the effective fracture is 24 m, and the entire length of the effective fracture is 48 m.

Figure 2.12 illustrates the cumulative effect of two nearby fractures overlapping:

- On the left, (a) represents a fractured zone generated by injection stimulation from the left side.
- In the right, (b) represents a fractured zone generated by injection stimulation from the right side.
- On the bottom of the figure, (c) represents the fractured zone, which is the summation of the two zones represented by (a) and (b).

In the center of (c), the summation of *DV* values from the tip of fractures on both the left and right sides will reach the critical value of 0.05 if the two zones are put in a proper neighborhood (i.e., with a proper value of stage interval), let the tip of the nearby fractured zones overlapping

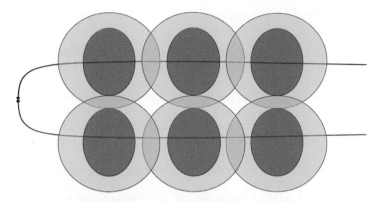

Figure 2.13   Cumulative effect of overlapping two nearby horizontal wells.

one another at the point represent half of the critical value of effective fracture. In this case, its *DV* value from each side is 0.025, and thus makes the sum a value of 0.05.

With reference to this analysis, the optimized stage interval will be twice the half-length of the fracture with its tip *DV* value as 0.025. The length of this value is 55 m; consequently, the optimized stage interval is 110 m.

With reference to the optimized stage interval of 110 m, the number of stages for this 400 m length of horizontal well is $N_s = 400/S = 400/110 = 3.65$, which is rounded to the closest larger integer, 4.

It is therefore suggested that four stimulation stages should be performed.

2.3.7   **Determination of the optimized well spacing based on numerical solutions**

The determination of the optimized well spacing is performed on the basis of the numerical solution shown in Figure 2.10 and Figure 2.11. This process is similar to the process described for the determination of the optimized stage interval. The major difference between the two processes is that the diagram shown in Figure 2.10 is used to determine the optimized stage interval, and the diagram shown in Figure 2.11 is used to determine the well spacing.

Figure 2.13 illustrates the cumulative effect of overlapping two nearby horizontal wells, showing a plane view of two parallel horizontal wells in the horizontal plane. The central area, shown in red, represents the effectively fractured area in which the *DV* value is greater than or equal to 0.05. The light green section beyond the central area has a *DV* value of less than 5%. Only microcracks are generated after hydraulic fracturing; consequently, this is a non-effective fractured region. The overlapping of two fractured zones increases the effective fractured area by the summation of *DV* values at their tips, as described in a previous section.

When well spacing is properly selected, the increased effective fracture zones will connect to one another between the two nearest stimulation stages from the two sides of the two wells.

As shown in the *DV* numerical solutions diagram (Fig. 2.11), the distance between the injection point and the tip point with a *DV* value of 0.025 is 63 m. Therefore, the optimized well spacing is set at twice the value of 63, which is 126 m. Therefore, the suggested well spacing of 126 m should be used when placing two parallel wells.

2.4   CONCLUSION

The major objective of this study is to develop a workflow with a real 3D fracture model. In this chapter, the 3D continuum damage model is adopted to model the fracture generation and propagation within a formation under hydraulic injection. The following list summarizes the conclusions drawn from this study:

- This work uses the 3D finite element method to calculate the distribution and evolution of the continuum *DV* under the given loading of injection flow rate. Fracture developments in each point of a given full 3D space are calculated.
- This work directly uses the numerical results of fracture development in two horizontal directions to design the stage interval and well spacing, respectively. The other existing methods, such the stress shadow method and/or "empirical method," cannot perform the fracture calculation in the lateral direction; consequently, they cannot use fracture development information for the stage interval design.
- This work uses the effective fracture concept as its criterion for the optimized design of stage intervals of multistage stimulation.
- This work uses a critical value of continuum *DV* at the front tip of the fracture zone as its criterion for determining the geometrical scope of an effective fracture.
- The determination of an optimized stage interval considers the overlapping effect of two nearby stimulation stages. This method ensures that the size of the optimized stage value is not too large to effectively connect one another to two stimulation stages (i.e., not too far apart from one another). It also ensures that the size of the optimized stage value is not too small, which would result in stimulation stages placed too closely together. This will ensure the maximized production rate and the maximized amount of unlocked petroleum reserves percentage at the target formation.
- The determination of optimized well spacing considers the overlapping effect of two nearby stimulation stages. This value of the optimized well spacing ensures that the two parallel wells are not too near or too far away from one another.

The stage interval of the multistage stimulation and the well spacing between nearby horizontal wells are two critical parameters in the hydraulic fracturing design for unconventional petroleum resources. The proper selection of values for these two parameters is a key component in obtaining successful oil and gas production from unconventional resources, such as tight-sand oil and shale gas. The method proposed in this work can accurately determine the optimized values for stage interval and well spacing. Accurate solutions of these two parameters ensure maximizing the oil and gas production rate and the percentage of unlocked petroleum resources reserved in the formations.

CHAPTER 3

Numerical analysis of the interaction between two zipper fracture wells using the continuum damage method

This chapter introduces the 3D numerical modeling of zipper fractures/modified zipper fractures. The purpose of this numerical modeling is to determine the most effective operational means to perform a hydraulic injection of the zipper fracture. With reference to the numerical damage distribution solution for a given injection plan, simultaneous injection was found to obtain a significantly larger area of damage distribution than that obtained by sequential hydraulic injection.

The plastic damage model is presented in Chapter 1; its parameter values, presented in Chapter 2, are adopted in this calculation.

3.1 INTRODUCTION

Zipper fracturing/modified zipper fracturing (Rafiee *et al.*, 2012) is a popular reservoir stimulation method used to develop unconventional resources, particularly for tight-sand oil and shale oil and gas. An adequate understanding of the influence of neighboring stimulation stages on the generation of the stimulated reservoir volume (*SRV*) significantly affects the fracturing design. To investigate the interaction mechanism between neighboring stimulation stages, numerical simulations were performed on the stimulation process. These simulations progress through each step of the stimulation process using a hydro-mechanical finite element method (FEM). A continuum damage model is used to simulate the fracture phenomena created by the fluid injection used for reservoir stimulation.

In rock mechanics, the mechanical damage variable is interpreted as an index of material continuity, which varies from 0 for intact rock to 1 for completely separated, broken rock. The volumetric density of cracks created by the injection fluid can be represented by a scalar damage variable.

Submodeling techniques are used to accommodate the field-to-borehole-section scale discrepancy. The submodeling technique uses a large-scale global model to produce boundary conditions for a smaller scale submodel. In this way, the hierarchical levels of the submodel are not limited. Using this approach, a highly inclusive field-scale analysis can be linked to a very detailed local-stress analysis at a much smaller scale. The benefits are bidirectional, with both the larger and smaller scale simulations benefiting from the linkage.

This chapter introduces a plasticity-based damage model and presents a set of connected hydro-mechanical problems calculated using a 3D model of the FEM. The numerical results for the distribution of the mechanical variables, including continuum damage, pore pressure, and horizontal stress components, are analyzed and shown.

The plastic damage model, presented in Chapter 1, and its values of parameters, presented in Chapter 2, are adopted in this calculation. The field model presented in Chapter 2 is used as the global model in this chapter. Section 3.2 presents the numerical results of the hydraulic fracturing simulation obtained with the submodel.

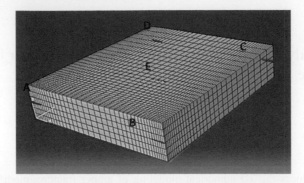

Figure 3.1   Mesh and geometry of the submodel.

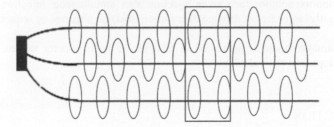

Figure 3.2   The parallel well horizontal sections designed for a modified zipper fracture.

3.2   SUBMODEL FOR STIMULATION PROCESS SIMULATION

For the submodel under stimulation injection loadings, the submodel geometry is established with a width and length of 200 m and a thickness of 20 m. The calculation focuses on the lateral scope of the fractured volume generated by the fluid injection. Therefore, the vertical direction of the model has been simplified by taking only a thin slice in 3D space.

The discretization of the model uses 7560 3D eight-noded continuum elements (C3D8RPH). Figure 3.1 shows the mesh of the submodel.

The load is the injection flow at the injection points of the submodel. Figure 3.2 shows the horizontal well sections designed the for zipper fracture. The red points shown in Figure 3.1 represent the locations of the injection points. Because of the symmetry of the model, only a quarter of the model was meshed. All four side-surfaces are symmetrical planes.

The task of this calculation is to estimate the injection effects with the previously described parameter values for the modified zipper fracture of two horizontal wells. The perforation length section is set at 20 m. Because of the symmetry of the geometric model, only 10 m of the perforation section are modeled with the injection points. The flow rate is the controlled loading variable and is provided as a known value; the pressure at the injection points is variable, solved as an unknown value, and changes with the injection process.

The numerical simulation of each injection step at the five locations (shown in Fig. 3.1) of the injection loading case are performed sequentially to simulate the stimulation process as practice for reservoir stimulation.

This calculation simulates fracture generation under a bottomhole pressure (*BHP*) of 39.5 MPa (5730 psi), along with the following conditions:

- The initial reservoir pore pressure is 20.7 MPa (3000 psi), corresponding to $TVD = 1700$ m.
- Boundary conditions are taken from the numerical results of the global model described previously.

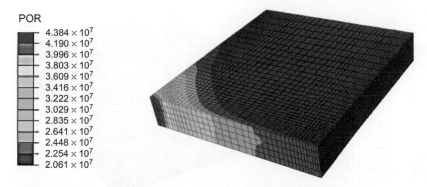

Figure 3.3   Pore pressure distribution of the submodel at Stage 1 of the injection stimulation.

Figure 3.4   Fractured volume at the Stage 1 distribution of continuum damage.

• The initial geostress values are taken from the numerical results of the global model described previously, also corresponding to $TVD = 1700$ m.

For the numerical results of the submodel under the stimulation injection loadings, the stimulation processes are numerically simulated.

To investigate the interaction of the fracture zones created by stimulation injection, the loading sequence used for the injection includes the following:

• Stage 1: injection loading at Location A
• Stage 2: injection loading at Location B
• Stage 3: injection loading at Location C
• Stage 4: injection loading at Location D
• Stage 5: injection loading at Location E

Figures 3.3 through 3.6 show the resulting contours of the major mechanical variables at the end of Stage 1 of the stimulation injection. Figure 3.3 shows the pore pressure distribution within the submodel corresponding to the end of the Stage 1 injection at Location A. Figure 3.4 shows the fractured reservoir volume, which is a rather narrow band in 3D space. Figure 3.5 shows the distribution of the horizontal stress component Sx in the X-direction. The stress components shown are all effective stresses, whose values are the result of the total stress minus the pore pressure. Figure 3.6 shows another horizontal stress component Sy in the Y-direction. The variation of the

Figure 3.5   Contour of the stress component Sx, which is the effective stress in the X-direction.

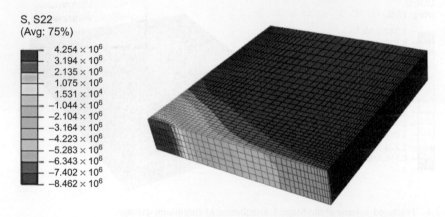

Figure 3.6   Contour of the stress component Sy, which is the effective stress in the Y-direction.

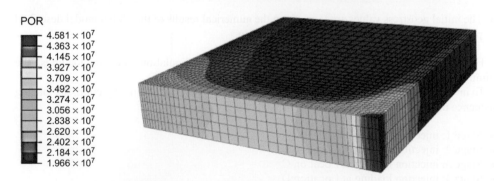

Figure 3.7   Pore pressure distribution of the submodel at Stage 2 of the injection stimulation.

vertical stress component Sz is the result of the amount of total vertical stress minus pore pressure, and is therefore omitted for brevity.

Figures 3.7 through 3.10 show the resulting contours of the major mechanical variables at the end of Stage 2 of the stimulation injection. Figure 3.7 shows the pore pressure distribution within the submodel that corresponds to the end of the Stage 2 injection at Location B. Figure 3.8 shows the fractured reservoir volume, which is also narrow. Although the pore

Figure 3.8   Fractured volume at the Stage 2 distribution of the continuum damage.

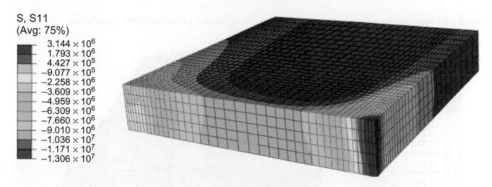

Figure 3.9   Contour of stress component Sx, which is the effective stress in the X-direction.

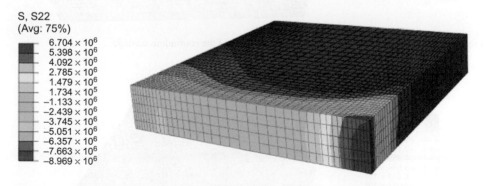

Figure 3.10   Contour of stress component Sy, which is the effective stress in the Y-direction.

pressure at Point A becomes increasingly less, its fractured volume and damage variable value continue to increase as a result of the residual injection energy at this region (which degrades with time). Figure 3.9 shows the distribution of the horizontal stress component Sx in the X-direction. Figure 3.10 shows another horizontal stress component Sy in the Y-direction.

Figures 3.11 through 3.14 show the resulting contours of the major mechanical variables at the end of Stage 3 of the stimulation injection. Figure 3.11 shows the pore pressure distribution in the submodel corresponding to the end of the Stage 3 injection at Location C. Figure 3.12 shows the fractured reservoir volume. Although it is still narrow, it connects the fracture volume created by injections at Location A and B, respectively, which means that the fractured volumes

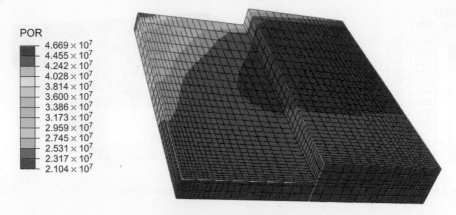

Figure 3.11   Pore pressure distribution of the submodel at Stage 3 of the injection stimulation.

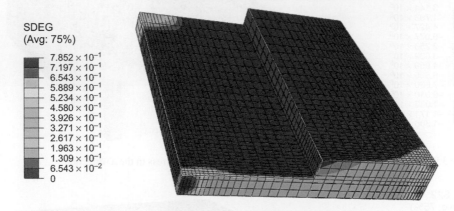

Figure 3.12   Fractured volume at the Stage 3 distribution of the continuum damage.

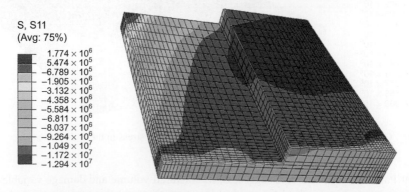

Figure 3.13   Contour of the stress component *Sx*, which is the effective stress in the *X*-direction (which is shown as S11 in legend).

are "zippered." Figure 3.13 shows distribution of the horizontal stress component *Sx* in the *X*-direction. Figure 3.14 shows another horizontal stress component *Sy* in the *Y*-direction. To clearly show the internal distribution of the mechanical variables, multi-cut views are used to visualize these figures.

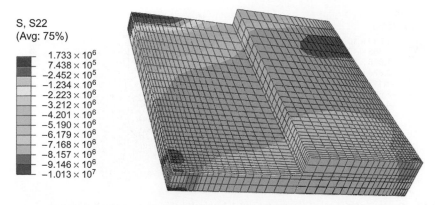

S, S22
(Avg: 75%)

1.733×10^6
7.438×10^5
-2.452×10^5
-1.234×10^6
-2.223×10^6
-3.212×10^6
-4.201×10^6
-5.190×10^6
-6.179×10^6
-7.168×10^6
-8.157×10^6
-9.146×10^6
-1.013×10^7

Figure 3.14   Contour of the stress component Sy, which is the effective stress in the Y-direction direction (which is shown as S22 in legend).

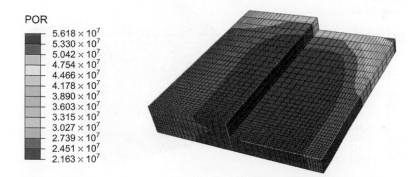

POR

5.618×10^7
5.330×10^7
5.042×10^7
4.754×10^7
4.466×10^7
4.178×10^7
3.890×10^7
3.603×10^7
3.315×10^7
3.027×10^7
2.739×10^7
2.451×10^7
2.163×10^7

Figure 3.15   Pore pressure distribution of the submodel at Stage 4 of the injection stimulation.

SDEG
(Avg: 75%)

8.102×10^{-1}
7.426×10^{-1}
6.751×10^{-1}
6.076×10^{-1}
5.401×10^{-1}
4.726×10^{-1}
4.051×10^{-1}
3.376×10^{-1}
2.701×10^{-1}
2.025×10^{-1}
1.350×10^{-1}
6.751×10^{-2}
0

Figure 3.16   Fractured volume at the Stage 4 distribution of the continuum damage.

Figures 3.15 through 3.18 show the resulting contours of the major mechanical variables at the end of Stage 4 of the stimulation injection. Figure 3.15 shows the pore pressure distribution within the submodel corresponding to the end of the Stage 4 injection at Location D. Figure 3.16 shows the fractured reservoir volume. All four fractured volumes are now connected or "zippered."

Figure 3.17   Contour of stress component *Sx*, which is the effective stress in the *X*-direction.

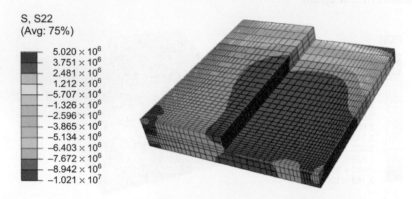

Figure 3.18   Contour of stress component *Sy*, which is the effective stress in the *Y*-direction.

Figure 3.19   Pore pressure distribution of the submodel at Stage 5 of the injection stimulation.

Figure 3.17 shows the distribution of the horizontal stress component *Sx* in the *X*-direction. Figure 3.18 shows another horizontal stress component *Sy* in the *Y*-direction.

Figures 3.19 through 3.22 show the resulting contours of the major mechanical variables at the end of Stage 5 of the stimulation injection. Figure 3.19 shows the pore pressure distribution within the submodel corresponding to the end of the Stage 5 injection at Location E. Figure 3.20 shows the fractured reservoir volume. The fractured reservoir volume created by this injection stimulation is significantly larger than that created by previous stimulation steps with approximately the same

Figure 3.20   Fractured volume at the Stage 5 distribution of the continuum damage.

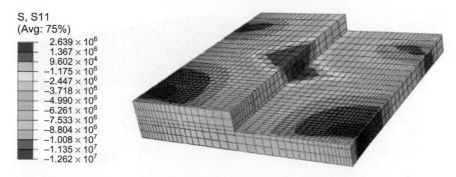

Figure 3.21   Contour of the stress component Sx, which is the effective stress in the X-direction.

Figure 3.22   Contour of the stress component Sy, which is the effective stress in the Y-direction.

injection pressure. Figure 3.21 shows the distribution of the horizontal stress component Sx in the X-direction. Figure 3.22 shows another horizontal stress component Sy in the Y-direction.

Figures 3.23 through 3.26 show the resulting contours of the major mechanical variables with a different stimulation injection method: two stimulation injections begin simultaneously at Locations A and B, rather than performed consecutively. Figure 3.23 shows the pore pressure distribution within the submodel corresponding to the simultaneous injection method. Figure 3.24 shows the fractured reservoir volume created in this case. The fractured volume is significantly wider than that shown in Figure 3.4 and Figure 3.10 with approximately the same injection pressure. This indicates that, for the zipper fracture under the given initial geostress, the same injection rate will produce more fractured reservoir volume using the simultaneous injection method than by using the sequential injection method. Figure 3.25 shows the distribution of the horizontal

POR

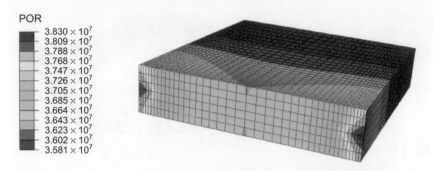

Figure 3.23   Pore pressure distribution of the submodel resulting from simultaneous injection stimulation.

SDEG
(Avg: 75%)

Figure 3.24   Fractured volume resulting from simultaneous injection stimulation, distribution of continuum damage.

S, S11
(Avg: 75%)

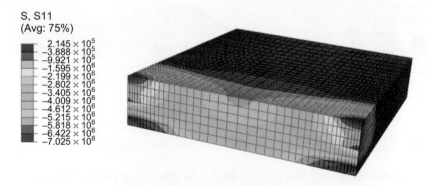

Figure 3.25   Contour of stress component Sx, which is the effective stress in the X-direction.

stress component Sx in the X-direction. Figure 3.26 shows another horizontal stress component Sy in the Y-direction.

3.3   CONCLUSIONS

The numerical results presented in this chapter include the following:

- Distribution of the fractures, which are represented by the contour of the continuum damage variable resulting from the injection flow.

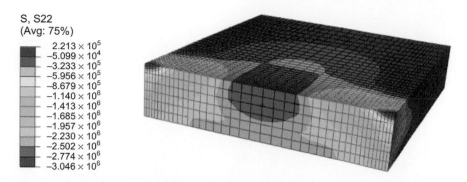

S, S22
(Avg: 75%)

2.213×10^5
-5.099×10^4
-3.233×10^5
-5.956×10^5
-8.679×10^5
-1.140×10^6
-1.413×10^6
-1.685×10^6
-1.957×10^6
-2.230×10^6
-2.502×10^6
-2.774×10^6
-3.046×10^6

Figure 3.26   Contour of stress component Sy, which is the effective stress in the Y-direction.

- Pore pressure distribution corresponding to the end of a given stimulation stage.
- Contours of the horizontal stress components Sx and Sy.

A comparison of the numerical results of *SRV* shown in Figure 3.4 and Figure 3.10 indicates that for a zipper fracture under the given initial geostress, the *SRV* generated by the sequential injection method is narrow and significantly less than that generated by the simultaneous injection method.

The numerical results of *SRV* shown in Figure 3.20 indicate that because the changes of the geostress field caused by neighboring injections, the *SRV* created by stimulation at the central area of this submodel model is much larger than those created at the corner locations.

As a special form of the zipper fracture, the so-called modified zipper fracture can increase efficiency of hydraulic fracturing operations.

The anisotropic distribution of permeability and distribution of the natural fractures are necessary data for obtaining a solution that is near the measured data of the fracture distribution obtained with microseismic monitoring. The calibration of the damage model with microseismic data will be a major part of the next step of this study. The calibration and modeling of the initial damage field is also important for accurately modeling this multi-physics phenomenon.

Figure 2.9 Contour of stress component S_1, which is the effective stress in the 1-direction.

- Pore pressure distribution corresponding to the end of a given simulation stage.
- Contour of the horizontal stress components S_2 and S_3.

A comparison of the numerical results of S_1, shown in Figure 2.9 and Figure 3.10 indicates that for a zipper fracture under the given initial geostress, the SRF, generated by the sequential injection method is narrow and significantly less than that generated by the simultaneous injection method.

The numerical results of S_1, shown in Figure 3.20 indicate that because the changes of the geostress field caused by neighboring interactions, the SRF created by stimulation at the central area of this submodel is much larger than those created at the corner locations.

As a special form of the zipper fracture, the so-called modified zipper fracture can increase efficiency of hydraulic fracturing operations.

The anisotropic distribution of permeability and distribution of the natural fractures are necessary data for obtaining a solution that is near the measured data of the fracture distribution obtained with microseismic monitoring. The calibration of the damage model with microseismic data will be a major part of the next version of this study. The calibration and modeling of the initial damage field is also important for accurately modeling this multi-physics phenomenon.

CHAPTER 4

Integrated workflow for feasibility study of cuttings reinjection based on 3D geomechanical analysis and case study

This chapter presents an integrated workflow for the feasibility study of cuttings reinjection (CRI) based on 3D geomechanics analysis. The contents of a CRI feasibility study include (i) selection of the well location; (ii) selection of the perforation section; (iii) 3D simulation of hydraulic fracturing; (iv) risk mitigation of the fault reactivation and fluid migration; (v) estimation of the magnitude of seismic activity.

Solutions of various mechanical variables obtained with a 3D geomechanics analysis at various levels of scale are used as the basis for designing CRI parameters. Solutions from geomechanics analyses provide the basis for a feasibility study and/or CRI design of the following: solution of 3D geostress distribution and the effective stress ratio are the essential factors for selecting the best location of injection well; solution of 1D geomechanics analysis provides a basis for selecting the true vertical depth (*TVD*) interval for injection sections; and hydraulic fracturing performed in the framework of 3D geomechanics analysis provides the most accurate solution for both the injection pressure window and fault reactivation related to CRI, as well as a seismic behavior estimation. An example of a CRI feasibility study with the proposed integrated workflow is presented with data from a case in offshore West Africa. Geomechanics analysis solutions are used for decision making at various CRI stages.

4.1   INTRODUCTION

Cuttings reinjection (CRI) is an engineering practice used for the disposal of drilling waste. During this process, hydraulic fractures are created at the target formation, and milled cuttings are injected with fluid. In practice, this process must comply with environmental regulations and zero-discharge policies. Because of the zero-discharge policy, fluid migration must be analyzed in the feasibility study and/or CRI design. Consequently, cap integrity and fault reactivation are two essential tasks to be performed in CRI, along with the hydraulic fracturing.

Among various accidents related to oil spill, the 2011 Bohai Bay oil spill is one caused by CRI (Wikimedia, 2012). This accident was caused by over-pressured CRI operation, and a big fault in the neighborhood of CRI operation section was reactivated, and consequently oil along with waste drilling fluid got spilled into sea water.

Geomechanical analysis solutions provide a basis for a feasibility study and/or design of CRI:

- The 3D geostress distribution and the effective stress ratio solution, which is defined in Section 4.4, are the essential factors used to select the most ideal location of the injection well.
- The 1D geomechanical analysis solution provides a basis for selecting a true vertical depth (*TVD*) interval for the injection sections.
- Hydraulic fracturing performed within the framework of the 3D geomechanical analysis provides the most accurate solution, for not only the injection pressure window, but also for the fault reactivation analysis related to CRI and seismic behavior estimates.

In the past decades, most of the CRI references focused on the operational aspects of this kind of design. The research reported by Reddoch *et al.* (1996) primarily introduces the work of well planning, surface devices, and injection pressure. The work of Chavez *et al.* (2006) simulated the fracture generated for cuttings reinjection and optimized the fracture geometry. The work reported in Bartko *et al.* (2009) has adopted geomechanical modeling and analysis in its CRI design process. A 1D geomechanical analysis was performed to select the *TVD* section of the injection point, along with the stress contrast method. Hydraulic fracture analysis was also performed in Bartko *et al.* (2009) with a 3D planar fracture model. In the work reported in Ezell *et al.* (2011), various operational aspects of CRI were discussed through a case performed in an offshore field in Saudi Arabia focusing on slurry properties. Geomechanical analysis was also used in Shen and Standifird (2015) for hydraulic fracturing simulation.

As one of the major tasks to be performed in the CRI analysis, the principle and techniques used in CRI hydraulic fracturing analysis are the same as those used in the analysis of reservoir stimulation for tight gas and tight oil (Bartko *et al.*, 2009; Soliman *et al.*, 2004; Warpinski, 2013).

Fault reactivation is an engineering problem that can be initiated by the connection of stimulated factures to the fault. Reactivation of a fault could result in environmental pollution, resulting from fluid migration, particularly in an offshore field. Another environmental risk of injection-related fault reactivation is the scenario of induced seismic behavior.

Fault reactivation related to hydraulic fracturing, either by CRI or reservoir stimulation, has been investigated by various researchers (Maxwell *et al.*, 2009; Pereira *et al.*, 2010; Soltanzadeh and Hawkes, 2007; Xu *et al.*, 2010). Analytical methods were used at the early ages, and the numerical method of 3D and/or quasi-3D became increasingly popular in recent years as a result of the development of computational technology.

Seismic analysis related to fault reactivation, as a result of hydraulic fracture related to either CRI or other purposes, was investigated by various researchers (Fatehi *et al.*, 2014; Hanks and Kanamori, 2008; Karvounis *et al.*, 2014; McGarr, 2014). Seismic analysis involving porous flow and 3D dynamic plastic behavior of faults is a complex process. Conversely, however, the purely analytical solution of a fault reactivation seismic analysis is rather too simple. Meshing parameters and numerical efficiency were investigated for hydraulic fracturing simulation (Gutierrez and Nygard, 2008; Sun and Schechter, 2015; Sun *et al.*, 2015).

Although the previously mentioned reference articles report various existing works, the following aspects in the current CRI practice can be improved:

- In the practice described in the literature cited and others, the selection of the injection well location is primarily based on geological information; an ideal geostress factor is not considered. An ideal stress factor, however, can significantly affect the success of a CRI operation (Sun and Schechter, 2014).
- In the cases reported in the references, such as Bartko *et al.* (2009) and Soliman *et al.* (2004), the focus of hydraulic fracturing analysis is on determining the values of surface pumping pressure and/or the injection rate, which will initiate the fracture. The upper bound of the injection pressure is not calculated.
- An estimate of the fault reactivation risk is not yet regarded as an essential part of a workflow for a feasibility study. This lack of information, however, is actually not in accordance with the "zero-discharge" environmental policy.
- There is not yet an integrated workflow that synthetically combines the solution of 3D geomechanical analysis with the major tasks of a CRI feasibility study.

In general, the proposed integrated workflow for a cuttings reinjection feasibility study should include the following seven steps:

1. Select the plane location of the injection well.
2. Select the *TVD* interval where the injection is applied.
3. Design the hydraulic fracturing procedure, which determines the proper injection rate and/or injection pressure.
4. Perform a cap integrity estimation, which verifies the safety of the injection pressure and injection rate under the constraint of the zero discharge policy.

5. Perform a fault reactivation analysis, which determines the generated fracture length under the constraints of the zero discharge policy.
6. Perform a seismic analysis of the fault reactivation.
7. Determine the volume of fluid with cuttings that can be injected at this well location.

This work will establish an integrated workflow for a CRI feasibility study. A solution of mechanical variables obtained with a 3D geomechanical analysis at different levels of scale will provide a solid foundation for this usage. Those problems previously described will be overcome with this workflow.

Section 4.2 provides details of the integrated workflow proposed for a cuttings reinjection feasibility study. A systematic CRI procedure will be established on the basis of 1D and 3D geomechanical analysis solutions to overcome the previously described shortcomings. 1D and 3D geomechanical analysis tools will be used as the major theoretical tools for this CRI feasibility study.

4.2   THE INTEGRATED WORKFLOW

Figure 4.1 shows the integrated workflow for a cuttings reinjection feasibility study. The CRI stages are shown in blue squares on the left side of the figure; the green squares on the right side show the associated geomechanical analysis tasks.

3D geomechanical analysis solutions have been used for decision-making at various CRI stages. The application of the 3D geomechanical analysis is one of the major characteristics of this proposed integrated workflow; the conventional 1D geomechanical analysis, however, is also a part of this workflow.

This integrated workflow consists of the following five steps:

1. Determine the location of the injection well. A 3D finite element calculation will be performed in this step. Stress distribution within the formation of the field will be calculated. A 3D solution of principal stress ratio, which is used as an index of ideal stress, will be calculated further from the solution of 3D stress distribution. An Abaqus User Subroutine is developed for this specific purpose.
2. Determine the *TVD* interval of the injection section. A 1D brittleness index solution will be calculated with the input of logging data, such as sonic data and gamma ray.
3. Calculate the injection pressure window (IPW). This step contains two parts: the value of the lower bound and the value of the upper bound of the IPW.
   a. The value of the lower bound of this IPW will be derived from the solution of the hydraulic fracturing 3D geomechanical analysis, which focuses on fracture initiation and propagation. This value is the injection pressure value at the stage of stable fracture propagation.
   b. The value of the upper bound of the IPW is the peak value of injection pressure with a proper injection rate. This value is also constrained by the cap integrity in general
   c. A 3D geomechanical analysis of the cap integrity is performed in this step to analyze the fluid migration, which is an essential part of the cap integrity verification.
4. Perform the fault reactivation analysis. In this step, a quasi-3D, which is actually 2D plane strain, finite element model is used for accuracy and efficiency. The first part of this step is a fault reactivation analysis to assess fluid migration. The second part of this step is an analysis of seismic behavior resulting from fault reactivation. The second part is also the estimate of the magnitude of seismic activity. A numerical solution of the displacement discrepancy across the fault will be calculated with a finite element model. However, the magnitude of the seismic activity related to the fault reaction will be calculated analytically.
5. Calculate the volume of the fracture generated. This is the value of fluid volume accommodated with cuttings to be injected. This volume is the product of the fracture's width, length, and height.

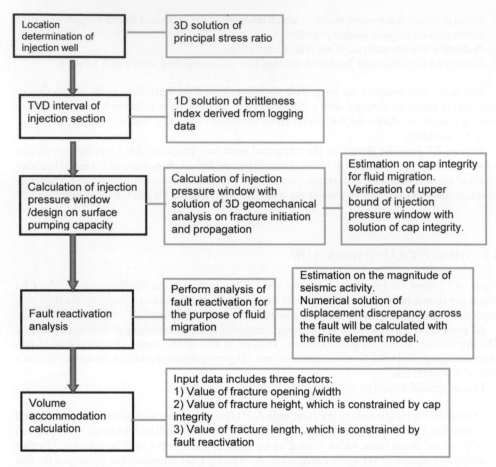

Figure 4.1   Flowchart of the integrated workflow for a CRI feasibility study based on 3D geomechanical analysis.

Additional details about the geomechanical analysis principles and concepts are introduced in the following sections.

The selection of the injection-well plane-location is based on the optimized value of the effective stress ratio. The principal stress ratio (PSR) is a ratio between the minimum principal effective stress (S_{min}) and maximum principal effective stress (S_{max}). Here, the sign convention of rock mechanics is adopted; i.e., compressive stress is positive, and tensile stress is negative. The definition of effective stress follows the definition from Terzaki's theory of porous elasticity. This PSR represents the discrepancy between these two principal stress components, and is expressed as:

$$r = |S_{min}/S_{max}|; \quad S_{max} \neq 0 \tag{4.1}$$

The adoption of this PSR is based on the experimental phenomena introduced in Gutierrez and Nygard (2008) and Sun and Schechter (2014). As shown in Figure 4.2, the ductile failure of shale rock occurs when the normal stress becomes high, and brittle failure occurs when normal stress becomes low. In solid mechanics, this normal stress is the third stress invariant (I_3), and the shear stress is a function of the second invariant (J_2) of the deviatoric stress tensor.

In reality, the smaller the value of γ, the greater the existing shear stress value will be in the initial geostress field, and consequently, the easier the rock formation fracturing will be.

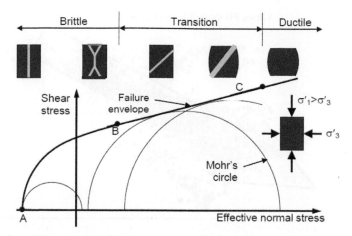

Figure 4.2   Brittle and ductile failure of shale rock with variation of shear stress (after Gutierrez and Nygard, 2008).

This means that the selection of the injection well should be at the region with the minimum value of γ.

In addition, the reason for not selecting a shear stress factor is that the calculation of the effective stress ratio in Equation (4.1) is much easier than that of a shear stress factor, which is represented by the stress invariant J_2. The final results for the location selection are near one another. Based on the cost-efficiency consideration, the current version of the effective stress ratio in Equation (4.1) is adopted as the index for selecting the location of the injection well.

To fulfill this task, the 3D finite element toolset Abaqus is used. A 3D elastoplastic stress analysis will be performed to obtain the 3D stress distribution on the nodes of the 3D field. The Abaqus User Subroutine technology is used to further calculate the effective stress ratio from the stress distribution solution for the given field.

Details of the finite element model for the purpose of this section are explained in Section 4.4 with the example of validation.

The selection of the *TVD* interval of the perforation section along a trajectory of the injection well is based on the brittleness index (*BI*). The *BI* is expressed in Equation (4.2):

$$BI = E/v \qquad (4.2)$$

With reference to the work introduced in Buller *et al.* (2010), the *BI* is proportional to the value of Young's modulus and is inversely proportional to the value of Poisson's ratio. Equation (4.2) has expressed the experimental phenomena shown in Figure 4.3.

In practice, the final determination of the *TVD* interval of the cuttings reinjection perforation section will be determined with the *TVD* cap interval. Cuttings reinjection requires the interval where the *BI* value is high, which is easy to fracture, and the cap interval needs the place where the value of *BI* is low, which is not easy to fracture.

To fulfill this task, the 1D analytical tool Drillworks® is used to derive the value of Young's modulus and Poisson's ratio from sonic logging data.

After the plane location of the well and *TVD* interval of the injection section are determined, the third task of this workflow is to perform a hydraulic fracturing analysis. This analysis includes two parts:

1. Determine the value of the injection rate, as well as the value of injection pressure, at bottom hole.
2. Perform the cap integrity analysis.

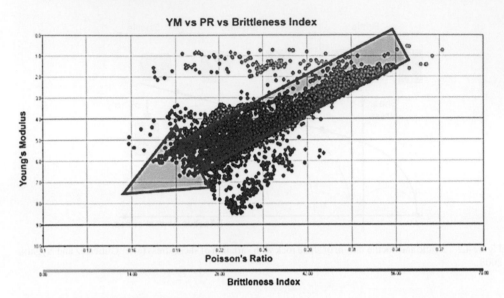

Figure 4.3   Illustration of brittleness index (after Buller *et al.*, 2010).

To fulfill this task efficiently and accurately, submodeling techniques have been adopted. Because of the accuracy requirement to capture the stress concentration around the fracture in a 3D plane, finer mesh is required for discretization of the model.

The submodeling technique concept includes using a large-scale global model to pro-duce boundary conditions for a smaller scale submodel. In this way, the hierarchical levels of the submodel are not limited. Using this approach, a highly inclusive field-scale analysis can be linked to a very detailed stress analysis at a much smaller borehole scale. The benefits are bidirectional, with both the larger and smaller scale simulations benefiting from the linkage.

Submodel 1 is designed for fracturing analysis in the horizontal direction, and Submodel 2 is designed for fracturing analysis in the vertical direction.

Three sets of injection pressure variation results will be obtained with these two submodels:

- For a given set of injection rate values, the curves of injection pressure *vs.* injection rate will indicate the lower bound of the IPW required to initiate a fracture, and the injection pressure value that must be maintained for fracture propagation.
- Value of the fracture opening/width for a given injection pressure value and injection rate.
- Upper bound of the IPW will be determined with numerical results of hydraulic fracturing. The peak values of injection of these two submodels will be compared.

Cap integrity analysis is performed on the basis of the previously described Submodel 2, which is designed for hydraulic fracturing simulation in the vertical direction.

This set of results will predict the fluid migration behavior and verify the upper bound of the IPW.

4.3   FAULT REACTIVATION ANALYSIS

The aims of fault reactivation analysis are dual and include the following:

- Estimation of fluid migration.
- Estimation of the maximum intensity level of seismic behavior.

In the following sections, these two aspects are described separately.

4.3.1 Fluid migration resulting from fault reactivation

The following assumptions and simplifications are adopted in the geomechanical modeling of this part:

- The plane-strain model is used for simplification purposes.
- The propagation process of injection-generated fractures within the formation other than a fault is neglected in this work. It is assumed that an injection-generated fracture has connected to the fault at one side, but will not cross the fault.

For accuracy and efficiency of the model, formations outside the fault area were taken as poro-elastic material, and fault material was modeled as poro-elastoplastic material.

The plastic strain-dependent permeability is adopted for the fault material, which means that the permeability of material of the fault will grow with the development of plastic strain. In this way, fluid migration will be modeled together with the development of the plastic deformation region. This is accomplished by using the Abaqus User Subroutines.

For accuracy and efficiency of the analysis, the fluid migration process is modeled as a transient process of porous flow.

The following numerical results of the mechanical variables are visualized:

- Distribution of plastic region, which shows the scope of fault being reactivated.
- Contour of pore pressure within the fault.
- Contour of von Mises equivalent stress and displacement field of the entire model.

With these modeling and numerical results, fluid migration scenarios related to the injection pressure will be simulated and predicted.

Details of the finite element model for the purpose of this section will be explained with the example of validation presented in Section 4.4.

4.3.2 Estimation of maximum intensity level of seismic behavior of the fault

In this work, the following techniques are used in the model:

- The level of magnitude of seismic activity resulting from fault reactivation is calculated analytically with an empirical equation.
- The input data of the calculation of magnitude of seismic activity includes the numerical solution of displacement discrepancy. This displacement discrepancy solution is obtained numerically with the finite element calculation of a simplified model.
- In this model, the Young's modulus values are assigned to each part of the model in a way that the resultant displacement discrepancy is localized to the region of the fault modeled in the analysis.
- The formation that consists of the upper side of the fault is modeled as "kinematic admissible." In this way, the model can simulate the kinematic behavior of a seismic activity
- For accuracy and efficiency of the seismic analysis, the porous flow occurring in the fault is regarded as the static porous flow process.

This analysis can predict the maximum level of magnitude of possible seismic activity.

4.4 EXAMPLES OF VALIDATION

This section presents an example of a cuttings reinjection feasibility study with the proposed integrated workflow. The geomechanical analysis solution is used for decision-making at various CRI stages. The data used in the numerical example in the following section is for workflow illustration purposes only. As shown in Figure 4.4, Field A0 is a major mature oil field in West Africa.

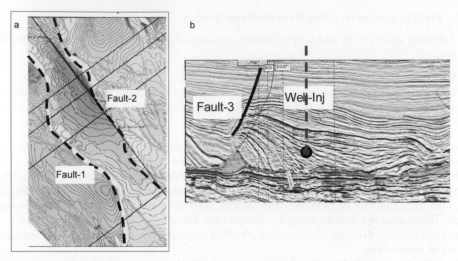

Figure 4.4 (a) Plane view of Field A0 with Fault 1 and Fault 2 in a NW–SE direction; (b) Fault 3 in a sectional view and the relative location of the candidate of injection well.

A well for cuttings disposal is planned in this field. The task of this work is to perform a feasibility study of this cuttings disposal well, including the following:

- Select the location of the injection well.
- Select the *TVD* interval of the perforation section for cuttings reinjection.
- Predict the *IPW*, along with the design of surface pumping capacity.
- Perform a fault reactivation analysis focusing on predicting the fluid migration and the maximum magnitude level of induced seismic activity.
- Predict the volume of fluid with cuttings that can be injected.

4.4.1 Location selection of the injection well

The purpose of 3D numerical modeling in this section is to determine the best location for the injection well. One of the selection criteria is to determine the location with the most ideal stress distribution, which will make the formation easy to fracture.

Figure 4.4 shows that salt formations exist below the formation investigated. Salt at the bottom significantly influences the distribution of the stress field; it applies high horizontal stress to the overburden formations.

4.4.2 Geometry and mesh

Figure 4.5 shows the geometry of the finite element method (FEM) model at the field level used in this work. The model dimensions are 12 km (length), 8 km (width), and 5 km (depth/thickness). Approximately 20,000 elements were used in the FEM mesh.

4.4.3 Values of material parameters

A poro-elastoplastic model was used in this work for the field model, and a poro-elastoplastic damage model was used for the submodel in the study of hydraulic fracturing in the following sections; a salt creep model is adopted for the salt formation. A variation of Young's modulus with the *TVD* is adopted. Table 4.1 lists the values of parameters of elasticity and density.

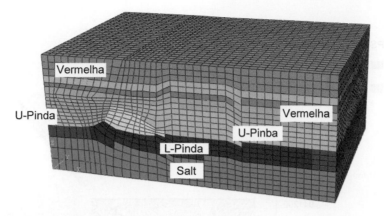

Figure 4.5   Geometry and mesh.

Table 4.1   Values of parameters of elasticity.

Formation	Poisson's ratio	Young's modulus [GPa]	Density [kg/m^3]
Top	0.3 to 0.31	6.353 to 6	2180
Lago	0.3 to 0.31	6.353 to 6	2180
Mesa	0.3 to 0.31	6.353 to 6	2180
Vermerha	0.3 to 0.31	6.353 to 6	2180
Upper Pinda (U-Pinda)	0.3 to 0.31	6.353 to 6	2180
Lower Pinda (L-Pinda)	0.3 to 0.31	6.353 to 6	2180

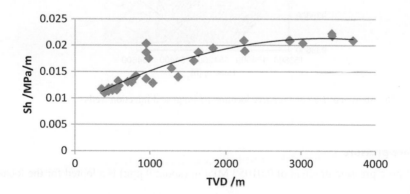

Figure 4.6   Leak-off test data for offset wells.

4.4.4   Initial geostress

The initial stress field was found as a normal fault stress pattern; vertical stress is the largest principal stress. Its direction of maximum horizontal stress is in the X-direction of the model. The initial value of the effective stress ratio is determined with reference to the leak-off test (LOT) data from offset wells in this field, and is shown in Figure 4.6. Figure 4.6 shows that the maximum value of minimum horizontal stress Sh reaches the value of 0.022 MPa/m (about 17 ppg), which indicates a principal stress ratio (*PSR*) value of approximately 0.8.

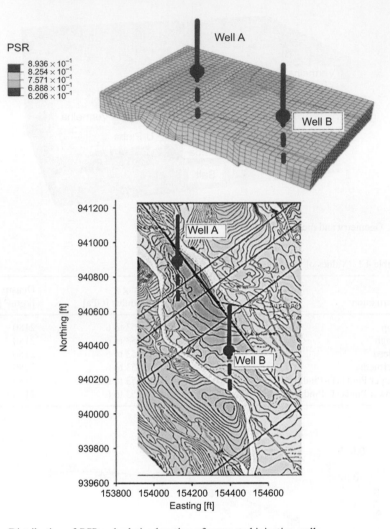

Figure 4.7   Distribution of *PSR* and relative location of suggested injection wells.

4.4.5   **Pore pressure**

A normal pore pressure gradient of 0.01059 MPa/m (about 9 ppg) is adopted for the location of injection.

4.4.6   **Numerical results of principal stress ratio**

Figures 4.7 and 4.8 provide the numerical results of the 3D stress analysis of the field. Figure 4.7 shows the distribution of the *PSR*. The smaller the value of *PSR*, the easier the hydraulic fracturing will be. Figure 4.8 shows the suggested locations of injection wells, taking the faulting factor into consideration. These two locations have the smallest values of *PSR* in the immediate area and are located far away from major faults.

In addition, with reference to the fracture simulation and fault reactivation, a fault reactivation risk exists if the fracture grows long enough and connects to any major fault. Consequently, it is necessary to locate the well target as far away as possible from faults.

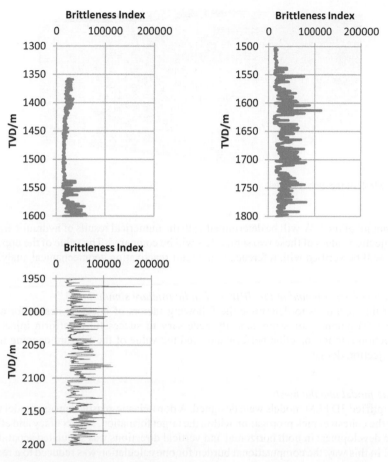

Figure 4.8   Distribution of brittleness index along well trajectory from $TVD = 1300$ to 2200 m.

Having considered all factors, the decision was made to locate the injection well (Well B) as shown in Figure 4.7.

In Figure 4.7, the top part of the figure is at the depth of the injection point, which is approximately $TVD = 2010$ m.

4.4.7   Selection of the true vertical depth interval of the perforation section

This injection well is designed as a vertical well. Figure 4.8 shows the BI distribution calculated with Equation (4.2) along the trajectory of the selected injection well. The logging data details and the Young's modulus and Poisson's ratio curves are neglected here for the purpose of brevity.

With reference to Figure 4.8, the *BI* value of within a section of 42 m (about 140 ft) *TVD* interval with its center at 2010 m is relatively greater than those in other intervals. Consequently, it is selected as the perforation section.

4.4.8   Fracture simulation: calculation of injection pressure window

The injection pressure window contains two parts: the value of lower bound and the value of upper bound. The value of the lower bound of this IPW will be derived from the 3D geomechanical analysis solution of hydraulic fracturing, which focuses on fracture initiation and propagation.

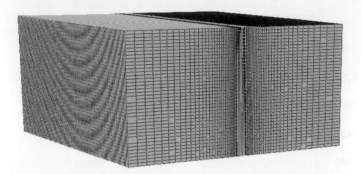

Figure 4.9   Mesh of the global model.

The upper bound of the IPW will be determined with the numerical results of hydraulic fracturing. The peak injection values of these two submodels will be compared. The value of the upper bound of the IPW will be verified with reference to the fault reactivation geomechanical analysis.

4.4.8.1   *Value of lower bound of this IPW: hydraulic fracture simulation*

The task of this section is to determine the following factors of injection design for a field in West Africa: (i) pumping pressure capacity necessary to successfully perform injection, and (ii) the injection rate for injection performance and the value of the fracture opening under the proposed injection design.

4.4.8.2   *The model and the mesh*

A set of simplified 3D FEM models were designed. A poro-elastic-plastic damage model was used to simulate the cohesive crack prorogation within the target formation. For accuracy and efficiency, the fracture development in both horizontal and vertical directions was simulated separately with submodels. In this way, the computational burden for one calculation was reduced to a reasonably low level without sacrificing model accuracy. The model loads included fluid injection and gravity, which balance the initial geostress field. Various values of the injection rate were used in these calculations to help determine the most reasonable injection pressure value. The pressure capacity of a pumping device was determined using this optimized injection pressure value.

The following paragraphs introduce model simplification. For accuracy and efficiency, the fracture development in the horizontal and vertical directions was simulated separately using submodeling techniques. The calculations of fracture propagation in a horizontal plane provide numerical solutions of these geomechanical variables: (i) fracture propagation situation, (ii) stress surrounding the fracture tip in the horizontal direction, and (iii) injection rate and pressure. The calculations of fracture propagation in a vertical plane provide solutions of fracture width as well as the injection rate and pressure.

Figure 4.9 shows the mesh of the global model of this calculation, which contains 320,000 elements. This model is used to calculate the stress field in the neighborhood of the target formation injection section. Non-zero displacement constraints have been applied to all surfaces of the model shown in Figure 4.9. The values of these displacement constraints are taken from the 3D model numerical results (shown in Fig. 4.6) used in Section 4.4.3 with the submodeling technique.

Submodel 1, shown in Figure 4.10, is designed for the calculation of the fracture initiation and propagation in the horizontal direction. Submodel 2, shown in Figure 4.11, is designed for the calculation of the fracture initiation and propagation in the vertical direction.

Table 4.2 shows the parameter values of damage-based fracture propagation. Cohesive element type is used for the fracture simulation. Quasi-brittle fracture behavior is assumed as the material model of the formation under hydraulic fracturing.

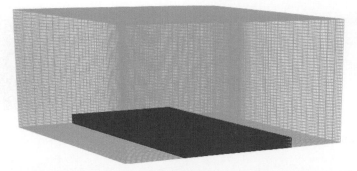

Figure 4.10   Geometry of Submodel 1 and its relative location in the global model.

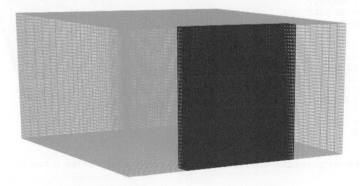

Figure 4.11   Geometry of Submodel 2 and its relative location in the global model.

Table 4.2   Values of parameters of damage-based fracture propagation.

Damage initiation, criterion = QUADS	$S1$ [Pa]	$S2$ [Pa]	$S3$ [Pa]
	1.2×10^6	5×10^6	5×10^6
Damage evolution, type = ENERGY	$G1$ [N/m]	$G2$ [N/m]	$G3$ [N/m]
	1000	20000	20000
Density [kg/m^3]	2180		
Elastic parameters	Kn [Pa]	$Kt1$ [Pa]	$Kt2$ [Pa]
	8.5×10^{10}	8.5×10^{10}	8.5×10^{10}
Fluid leak-off/gap flow	k-bottom [m/s]	k-top [m/s]	Kg [m/s]
	1×10^{-10}	1×10^{-10}	1×10^{-10}

In Table 4.2, $S1$ is the strength value in the normal direction of the fracture surface. $S2$ and $S3$ are strength values in the two directions tangential to the fracture surface. $G1$, $G2$, and $G3$ are the values of fracture energy for each single-mode fracture in the directions normal and tangential to the fracture surface, respectively. Kn, $Kt1$, and $Kt2$ are interfacial stiffness in the directions normal and tangential to the fracture surface, respectively.

4.4.8.3   *The load of fluid injection*
Figure 4.12 shows the geometry and mesh of Submodel 1 for horizontal fracture propagation. The geometry of the model is defined with 120 m in length, 100 m in width, and 10 m in thickness. Loads are defined with the gravity (initial geostress) and injection flow, as shown in the figure.

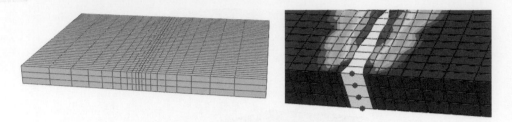

Figure 4.12   Geometry and mesh of the numerical model.

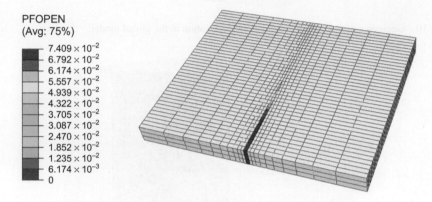

Figure 4.13   Fracture propagation and fracture opening for a given injection rate at time t.

4.4.8.4   *Numerical results of Submodel 1*

Figure 4.13 shows the fracture propagation status for a given injection rate at a time moment t. It shows the value of the fracture opening at a state of stable fracture propagation generated by cuttings disposal injection. Figure 4.14 shows the stress distribution result around the fracture as it develops in the horizontal direction. Figure 4.15 shows the comparison between the results of injection pressure given two different injection rates.

With reference to the numerical results shown in Figures 4.12 through 4.14, information about fracture width and length includes the following:

- The maximum fracture width is 74 mm.
- The fracture length can grow as long as injection continues before reaching a fault.

The rate and pressure are given for injection at $TVD = 2050$ m. The following values of injection rate and pressure are used:

- Rate 1 = 0.5616 m^3/min (3.6 bbl/min).
- Rate 2 = 0.05616 m^3/min (0.36 bbl/min).
- Rate 3 = 2.18 × 10^{-4} m^3/min (0.0014 bbl/min).

The numerical results obtained with Rate 2 indicate that maximum bottomhole pressure (*BHP*)/injection pressure is approximately 80 MPa; it stabilizes at approximately 46 MPa after the primary peak value.

For various injection rates, Figure 4.16 indicates that the maximum value of *BHP* varies, but the stable value of *BHP* for fracture propagation changes very little.

The injection rate is on the 10 m length section. The total rate of the injection volume can be obtained by calculating with the total length of the perforation section if it is not 10 m.

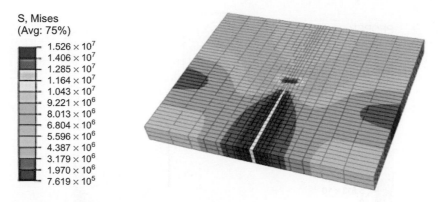

Figure 4.14   Result of stress distribution around fracture at time *t*.

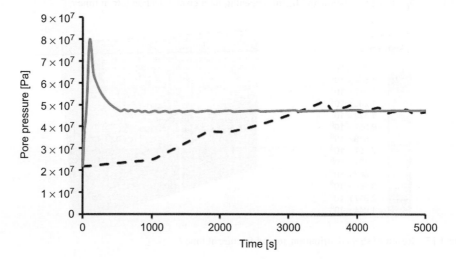

Figure 4.15   Comparison between results of injection pressure given two different injection rates.

4.4.8.5   *Numerical results of Submodel 2*

Figure 4.16 shows the fracture propagation of the vertical fracture model. From the numerical results of the vertical fracture propagation, the fracture width was determined to be 69.7 mm. The length of the fracture can grow to cap the formation with the given injection rate value.

Figure 4.17 shows the distribution of von Mises stress during the fracture propagation process. Figure 4.18 shows the variation of injection pressure *vs.* time at an injection rate of $0.05616 \, \text{m}^3/\text{min}$ (0.36 bbl/min). The maximum *BHP* is 60 MPa, and its stable value for fracture upward growth is 45.5 MPa.

During injection, the horizontal (H) and vertical (V) fractures will grow simultaneously, but at different rates. Figure 4.18 shows a comparison between injection pressure variation and time for horizontal and vertical fractures, respectively.

In summary, the following conclusions are reached:

- The lower bound of the IPW is determined to be 46.6 MPa.
- The fracture opening is determined to be 0.069 m.
- The upper bound of the IPW is determined to be 80 MPa.

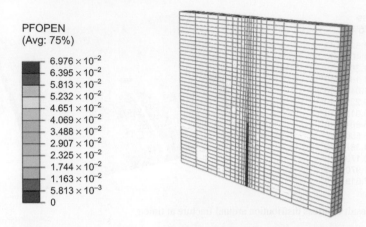

Figure 4.16   Fracture propagation and fracture opening for a given injection rate at time t.

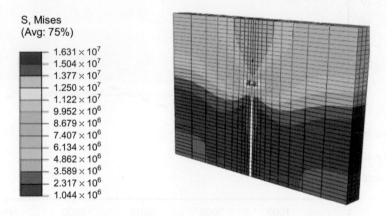

Figure 4.17   Result of stress distribution around fracture at time t.

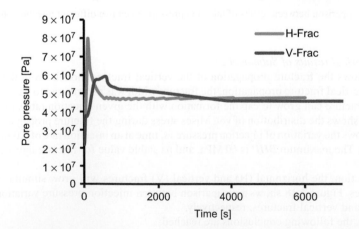

Figure 4.18   Injection pressure variation *vs*. time for horizontal and vertical fractures, respectively.

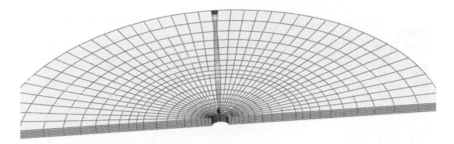

Figure 4.19   Crack propagation in the near wellbore field.

Anisotropy of the formation property was not considered in this work because of the lack of experimental and measured data, such as sonic logging and seismic data. Anisotropy must be considered for a more accurate estimate with respect to fracture propagation in two directions.

4.4.8.6   *Discussion on the fracture width value*

The fracture width value of 0.069 m previously described is obtained with the flow rate of 0.05616 m^3/min (0.36 bbl/min) with a 10 m length of the perforation section. Its peak injection pressure value is 80 MPa, and the stable injection pressure value is 46 MPa.

As discussed in Guo *et al.* (2000) and Willson *et al.* (1993), as a result of the fracture mechanics analysis, the shape of a fracture created by CRI can be elliptical, and the maximum fracture width value can be 0.0254 m and 0.0508 m, respectively. This fracture width value depends on the injection rate value.

If the variation of the stress values with *TVD* is neglected, the maximum fracture width value occurs at the location near the center of the fracture, and the width value becomes smaller at the top tip and bottom tip for a vertical fracture.

If the variation of stress values with *TVD* is considered, because the value of horizontal stress is smaller in the upper part and greater in the lower part (for the same injection rate and the same injection pressure), the crack width value with be larger in the upper part and smaller in the lower part.

This fracture width value is obtained with the model without the shape of the borehole; consequently, it is the resultant fracture width for fracture propagation in the far field. It is not the fracture opening at the surface of the wellbore. Figure 4.19 shows a fracture shape in the horizontal plane for the fracture propagation near the borehole. As a result of the stress concentration around borehole, the hoop stress value near the borehole surface is much larger than the stress value away from the borehole surface. Consequently, the fracture width at the borehole surface is much smaller than the value for a point away from the borehole. Because the fracture width value at the borehole surface is not a concern of the current study, its precise value is neglected here.

In general, if the perforation section is located in the lower part of the formation for CRI, this fracture width value is acceptable to use for the calculation of the accommodation volume. If the location of the perforation section is in the upper part of the CRI target formation, the accommodation volume could be somehow smaller than the value calculated, as shown at the end of the chapter.

4.4.8.7   *Cap integrity estimation*

This subsection demonstrates the process used to estimate the cap integrity; here, only the cap integrity is considered in the vertical direction.

Figure 4.20 shows Submodel 2, which is the FEM model for vertical fracture analysis. It is adopted to estimate the cap formation. The cap formation has been added at the upper part of the model. As shown in Figure 4.20, for simplicity, the cap formation directly sits 20 m above the injection formation. Table 4.3 provides the fracture property parameter values of the cap

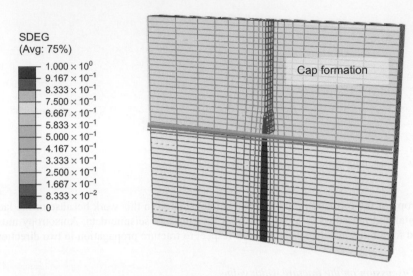

Figure 4.20   Finite element method model for cap integrity estimation and numerical results.

Table 4.3   Values of parameters of fracture property of the cap formation.

CDM-based fracture property			
Damage initiation, criterion = MAXS	S1 [Pa]	S2 [Pa]	S3 [Pa]
	2.4×10^6	10×10^6	10×10^6
Damage evolution, type = ENERGY	G1 [N/m]	G2 [N/m]	G3 [N/m]
	0.01×10^6	0.2×10^6	0.2×10^6

formation. As shown in the table, the values for the strength and fracture properties of the cap formation have been assigned as twice the value of that for the normal formation injected.

Figure 4.21 shows the variation of injection pressure with the existence of the cap formation. The value of stable injection pressure *BHP* has increased as a result of the increase of fracture energy (G_1). Because the cap formation is located some distance away from the injection point, the fracture initiation pressure does not increase.

Figure 4.22 shows the variations of parameter values G, v, and c at the cap formation, as well as the variation of the curves of injection pressure *vs.* time for the stable fracture propagation stage. Those details are not shown here.

In addition to the diagrams shown in Figure 4.21 and Figure 4.22, the numerical results of fracture propagation were verified. The fracture propagation was found to stop at the bottom of the cap formation, which ensures the integrity of the cap.

Another way to estimate the cap integrity under injection pressure is to simulate the case where the injection point is set at the location of the cap formation bottom.

Figure 4.23 shows the numerical results of this calculation. The pressure value required to initiate a fracture at the cap formation is significantly greater than the pressure values at the stage of stable fracture propagation at the injection perforation. This result confirms the integrity of the cap formation for this set of injection pressure values and mechanical properties of cap formation.

4.4.8.8   *Analysis of fault reactivation*
In this section, a 2D plane strain finite element model is used. The first part of this section demonstrates a fault reactivation analysis to assess fluid migration. The second part illustrates the analysis of seismic behavior resulting from fault reactivation and performs an estimation on

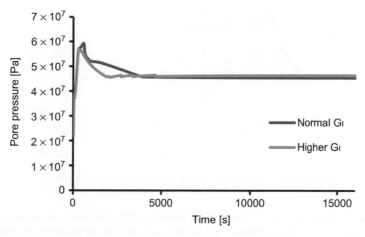

Figure 4.21   Variation of injection pressure *vs.* time: a comparison of two curves for with and without cap formation.

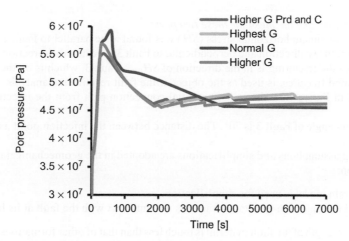

Figure 4.22   Comparison of curves of variation of injection pressure *vs.* time for stable fracture propagation stage.

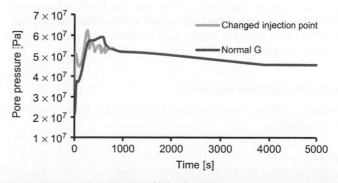

Figure 4.23   Comparison of curves of variation of injection pressure *vs.* time for stable fracture propagation stage.

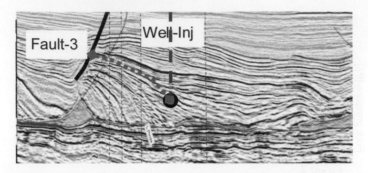

Figure 4.24   Fault 3 in a sectional view and the relative location of the injection well.

the magnitude of seismic activity. The numerical solution of displacement discrepancy across the fault is calculated with the finite element model and used as input to calculate the magnitude of seismicity. The magnitude of the seismic activity related to the fault reaction will be calculated by an analytical equation provided in Section 4.5.

4.4.8.9   *Assumptions and the mesh for these two parts*
The direction of maximum horizontal stress (SH) was found to be parallel to Fault 1 and Fault 2 in a northwest-southeast direction and perpendicular to Fault 3. Because the direction of fractures induced by hydraulic fracturing is in the direction of $SH_{\max}$, Fault 3, which is on the path of the injection-generated fractures, is used as the object of the fault reactivation analysis. The dashed dark green line in Figure 4.24 shows the potential connection path from the injection point to Fault 3.

The inclination angle of Fault 3 is 70°. The distance between the injection point and Fault 3 is 2000 m (6561 ft).

The following assumptions and simplifications are adopted in the geomechanical modeling of fault reactivation:

- The plane-strain model is used for simplification.
- It is assumed that injection-generated fractures will intersect with the fault at its bottom.

The cohesive strength of the fault material is much less than that of other formations. Therefore, a porous elastoplastic material model is adopted for the fault, and an elastic material model is adopted for other regions of the model.

Figure 4.25 shows the model geometry and FEM mesh used in the calculation.

The geomechanical model of fault reactivation used the following geometrical parameter values: height (H) = 1000 m, length (L) = 1500 m, and width of fault (B) = 20 m.

For the discretization of the model in Figure 4.26, the first-order plane-strain element CPE4R and pore-pressure stress coupled element CPE4RP were adopted.

4.4.9   **Fault reactivation and fluid migration**

Three types of formation lithology were included in this model: Fault, Upper Pinda formation, and Lower Pinda formation.

4.4.9.1   *Material data*
The mechanical behavior of the model used to analyze fluid migration is transient static, which means that the deformation behavior of the formation matrix is static. Consequently, Young's modulus is adopted as the static value. The porous flow of the model is transient. A variation of Young's modulus with depth was adopted. Table 4.4 shows the values for the elasticity parameters

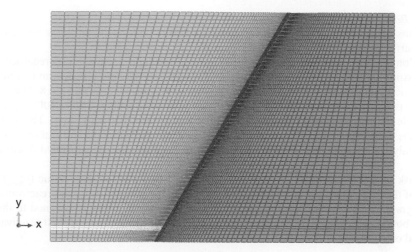

Figure 4.25   Model geometry and finite element method mesh.

Table 4.4   Values for parameters of elasticity.

Formation	Poisson's ratio	Young's modulus [GPa]	Density [kg/m^3]
Fault	0.24 to 0.31	3 to 7	2180
Upper Pinda formation	0.25 to 0.31	4 to 7	2180
Lower Pinda formation	0.25 to 0.31	4 to 7	2180

used in the model. The Young's modulus values for the seismic behavior analysis are introduced in Section 4.5.

The strength parameter values of the fault are given as cohesive strength (C) where $C = 1$ MPa and frictional angle (φ) where $\varphi = 25°$.

4.4.9.2   *Initial geostress*
This section provides the values of the initial geostress field used in the model.

The total stress value in the vertical direction is $\sigma_y = -32.3$ MPa at the bottom layer, where a value of coordinate $y = 0$ m; $\sigma_y = -11$ MPa at the top layer of the model, where $y = 1000$ m.

The sign convention of solid mechanics is adopted, with which compressive stress is negative.

With reference to the LOT data from offset wells in this field, the lateral stress coefficient was set as 0.8 (i.e., $\sigma_x = 0.8\sigma_y$).

4.4.9.3   *Pore pressure*
A normal pore-pressure gradient of 0.001 MPa/m was adopted for the injection location of this model, and its pressure value was set as a linear variation with *TVD* in the following form:

$$p_0 = (1000 - y)a + b \tag{4.3}$$

where $a = 10$ MPa/1000, $b = 10.8$ MPa, and y is the value of the Y-coordinate of a node in the mesh.

4.4.9.4   *Loads*
A gravity load was applied to the model along with overburden pressure on the top of the model, which represents the overburden applied on the top surface of the model caused by seawater.

4.4.9.5   *Boundary conditions*

Constraints of zero-normal displacements were applied to both lateral sides, as well as to the bottom surface of the model.

Injection pressure, which is the major driving force for fault reactivation, is regarded as the pore pressure boundary condition. The value of this injection pressure was determined by using the numerical analysis of hydraulic fracturing described previously in section 4.4.8. The value of 37.8 MPa is assumed for the pore pressure at the bottom of the fault at $TVD = 1000$ m, with the bottomhole pressure at the injection point of 46.6 MPa at $TVD = 2050$ m (6700 ft). This value was used as the maximum pore pressure, which was applied to the fault bottom location. This case applies only when a complete connection occurs between hydraulically generated fractures and the fault.

The task here was to numerically determine the maximum pore pressure value (p^s_{max}) at the bottom of the fault that could be sustained without resulting in significant fault reactivation and fluid migration. In the following section, a set of pore pressure values is assigned to the pore pressure boundary condition at the bottom of the fault, and the probability of fault reactivation is examined on a case-by-case basis.

4.4.9.6   *Numerical results of fault reactivation and fluid migration*

Cases of various pore pressure boundary conditions are given as the following:

- Case 1: $p1 = 25$ MPa.
- Case 2: $p2 = 35$ MPa.
- Case 3: $p3 = 37.8$ MPa.

The following numerical results of mechanical variables are visualized:

- The distribution of the plastic region, which shows the scope of the fault being reactivated for all three cases.
- The initial value of pore pressure field is assumed to be the hydrostatic pressure.
- The contour of pore pressure within the fault for Case 1.
- The contour of von Mises equivalent stress and displacement field of the whole model for the Case 2.

1)   Case 1: $p1 = 25$ MPa

Figures 4.26 and 4.27 show the contour of activated plastic status index of active yielding (AC Yield) and that of pore pressure for the beginning stage of fault reactivation. As shown Figure 4.26, the fault was activated in a small scope of range at two ends (indicated in red) when the injection fluid reached the bottom of the fault with a pore pressure of $PP = 25$ MPa. Each figure contains a legend explaining the color code that indicates the values for the figure.

2)   Case 2: $p2 = 35$ MPa

Omitting the evolutionary process of the mechanical variables, Figure 4.28 shows the contour of activated plastic status index AC for Case 2. As shown in Figure 4.28, the sections of the fault being activated grow from the two ends toward the central part under the pore pressure of 35 MPa. In this case, the injection-induced crack is almost completely connected with the fault.

3)   Case 3: $p3 = 37.8$ MPa

For this case, the numerical solution for the contour of activated plastic status is shown as index AC for Case 3. Figure 4.29 shows that the fault is reactivated along its entire length/height. In this case, the injection-induced crack is completely connected with the fault, and the pore pressure under the bottom of the fault reaches its maximum value of 37.8 MPa.

Figure 4.30 shows the contour of von Mises stress, and Figure 4.31 shows the contour of displacement magnitude for Case 3 as $PP = 37.8$ MPa.

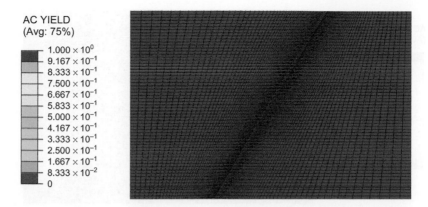

Figure 4.26   The fault was activated in a small scope of range at two ends under $PP = 25$ MPa.

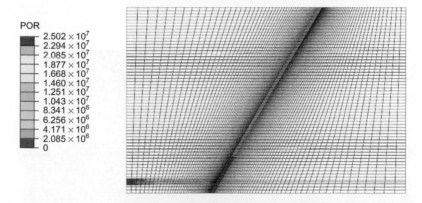

Figure 4.27   Distribution of pore pressure for the beginning stage of fault reactivation.

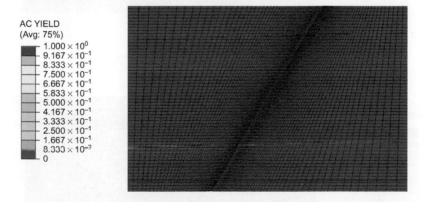

Figure 4.28   Contour of activated plastic status index AC for Case 2.

4.5   FAULT REACTIVATION AND SEISMICITY ANALYSIS

4.5.1   Analytical equation used to calculate the magnitude of seismic activity

Seismicity analysis related to fault reactivation caused by pore pressure variations has been investigated by various researchers. Equations (4.4) and (4.5) provide for the analytical solution of

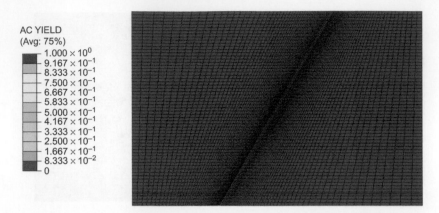

Figure 4.29   Contour of activated plastic status index *AC* for Case 3.

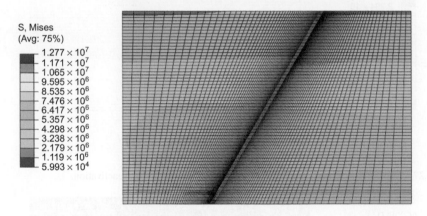

Figure 4.30   Contour of von Mises stress for Case 3.

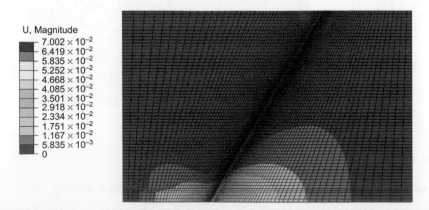

Figure 4.31   Contour of displacement magnitude for Case 3.

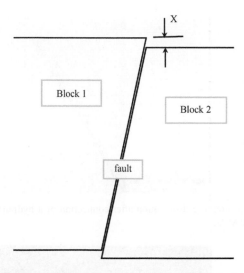

Figure 4.32   Illustration of relative displacement of two surfaces across a fault.

magnitude of seismicity that was empirically proposed by Hanks and Kanamori (1979), and subsequently used by other researchers (Fatehi *et al.*, 2014; Kanamori and Brodsky, 2001; Karvounis *et al.*, 2014; Mortezaei and Vahedifard, 2014):

$$M_0 = GXA_f \tag{4.4}$$

where, G is the shear modulus of the fault material, X is the average value of relative displacement across the fault, and A_f is the area of the fault surface experiencing the reactivation. This work uses the numerical solution of displacement discrepancy obtained with the finite element model described in the following Section 4.5.3 as the value of X. The shape of the sliding area at the fault surface is assumed to be circular. A convenient measure of the magnitude (M) of an earthquake resulting from fault reactivation is given in terms of the seismic moment by:

$$M = \frac{2}{3}\left[\log_{10}(M_0) - \alpha_0\right] \tag{4.5}$$

In Equation (4.5), the constant α_0 is a parameter determined by the empirical measurement of the region investigated. Hanks and Kanamori (1979) and Kanamori and Brodsky (2001) recommended a value of 9.1, but was given as 9.05 by McGarr (2014). This difference indicates that the parameter α should be calibrated with the measured seismicity data of the region investigated. Here, it is given the value of 9.1 as its default value.

Figure 4.32 illustrates the relative displacement (X) of two surfaces across a fault. To achieve such a relative displacement, a kinematic admissible field should be included in the numerical model. Although the model proposed in the previous sections is sufficient for fault reactivation purposes, which primarily targets fault deformation and fluid migration, it is not sufficient for seismic analysis. For seismic analysis, it is necessary to improve the previous model and enable it to include a kinematic admissible displacement field.

4.5.2   Assumptions and simplifications adopted in the finite element method

To simulate the seismic behavior of the model, the following changes were made to the model used here:

- The Young's modulus value of both the Lower Pinda and Upper Pinda formations was changed to a large value (1000 times greater than its original value listed in Table 4.4), but the Young's

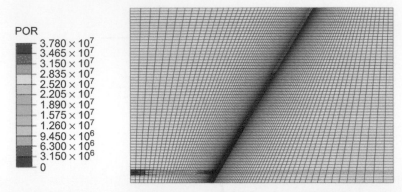

POR
- 3.780×10^7
- 3.465×10^7
- 3.150×10^7
- 2.835×10^7
- 2.520×10^7
- 2.205×10^7
- 1.890×10^7
- 1.575×10^7
- 1.260×10^7
- 9.450×10^6
- 6.300×10^6
- 3.150×10^6
- 0

Figure 4.33   Contour of pore pressure distribution after connection of a hydraulically generated fracture with the fault with $PP = 37.8$ MPa.

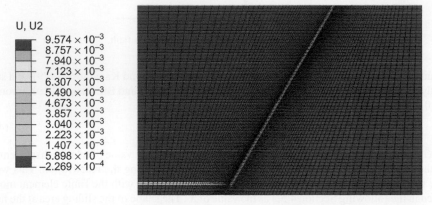

U, U2
- 9.574×10^{-3}
- 8.757×10^{-3}
- 7.940×10^{-3}
- 7.123×10^{-3}
- 6.307×10^{-3}
- 5.490×10^{-3}
- 4.673×10^{-3}
- 3.857×10^{-3}
- 3.040×10^{-3}
- 2.223×10^{-3}
- 1.407×10^{-3}
- 5.898×10^{-4}
- -2.269×10^{-4}

Figure 4.34   Contour of numerical results of the vertical displacement component for Case 1 with $PP = 30$ MPa.

modulus value at the fault used the same value as previously used. Consequently, the deformation energy of the entire model will either be dissipated by or stored in the fault through deformation.

- A weak layer was added to the model at the bottom of the Upper Pinda formation, which makes the formation relatively "kinematic admissible"; constraints from the surrounding formations become weaker than they were originally.

4.5.3   Numerical results

Two cases of displacement calculation for the purpose of seismic behavior analysis are performed with $PP = 30$ MPa as Case 1 and $PP = 37.8$ MPa as Case 2.

Figure 4.33 shows the contour of the pore pressure distribution after the connection of a hydraulically generated fracture with the fault at its lower end. The lower color zone is the location of the weaker formation of the model. The maximum value of pore pressure was set as 37.8 MPa.

From Figures 4.34 and 4.35, the displacement discrepancy value for Case 1 of 0.0096 m is obtained, and the value for Case 2 is 0.0286 m.

In this calculation, parameter values are:

$$G = 3 \times 10^9 \text{ Pa} \quad X = 0.0286 \text{ m}; \quad A_f = 7.85 \times 10^5 \text{ m}^2$$

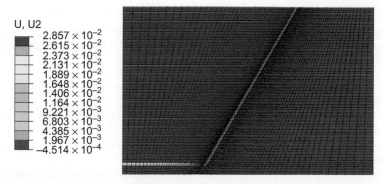

U, U2
2.857 × 10^{-2}
2.615 × 10^{-2}
2.373 × 10^{-2}
2.131 × 10^{-2}
1.889 × 10^{-2}
1.648 × 10^{-2}
1.406 × 10^{-2}
1.164 × 10^{-2}
9.221 × 10^{-3}
6.803 × 10^{-3}
4.385 × 10^{-3}
1.967 × 10^{-3}
−4.514 × 10^{-4}

Figure 4.35   Contour of numerical results of the vertical displacement component for Case 1 with $PP = 37.8$ MPa.

Consequently,

$$M_0 = GXA_f = 6.71 \times 10^{13}$$

Furthermore, the magnitude of the seismicity corresponding to the fault reactivation is as follows:

$$M = \frac{2}{3} \left[\log_{10}(M_0) - \alpha_0 \right] = \frac{2}{3}[13.826 - 9.1] = 3.15$$

Therefore, the maximum seismic magnitude caused by the fault reactivation of Fault 3 can reach the level of 3.15.

With reference to Equations (4.4) and (4.5), the intensity of seismicity depends on the amount of relative displacement across the fault. In addition, the magnitude of the relative displacement depends on the synthetic effect of the following two factors: (i) the existence of weak layers, which provides a kinematic admissible field of displacement constraints and (ii) the pore pressure value. Furthermore, abnormally high shear stress, along with high tectonic stress, also affects the seismic behavior of a reactivated fault.

Therefore, for an accurate analysis of the seismic behavior of fault reactivation, it is essential to have a thorough knowledge of the spatial structure around the fault.

Because the tectonic stress in this area belongs to a normal stress pattern, the magnitude of seismic behavior of the fault being activated is not significant.

4.5.4   Remarks

A numerical geomechanical model was built for the fault reactivation analysis in the neighborhood of a well drilled for cuttings reinjection. Two kinds of analyses were performed for fault reactivation: the first analyzes fluid migration, and the second analyzes seismic behavior. The following conclusions can be obtained from the numerical results:

- A pore pressure of 37.8 MPa, which is the value of the bottomhole pressure at the injection point of the cuttings reinjection operation, will likely cause significant reactivation of Fault 3 in the field. The injection fluid will migrate to the top of the fault.
- When the pore pressure reaches $p2 = 35$ MPa at the fault, fault reactivation is also significantly more likely. In this case, hydraulically generated fractures have begun to connect with the fault, but are not yet fully connected.
- It was necessary in the design of the cuttings reinjection to maintain the generated fracture length to a value of less than 2000 m, which was the distance from the injection point to the fault.

Seismicity properties were analyzed using simplified theoretical equations. From the numerical results, the calculated magnitude of fault reactivation would reach a maximum level of a magnitude 3.15 earthquake.

4.5.5 **Prediction of the volume of fluid with cuttings that can be injected**

In this section, the calculation of the generated fracture volume will be performed. This value is the volume of fluid accommodated with cuttings to be injected. This volume equals the product of the fracture's width, length, and height:

- With reference to the numerical result obtained in Section 4.4.8, the fracture width is $W = 0.069$ m.
- With reference to the fault reactivation analysis, the fracture length must be limited to less than the value of distance between the well and fault, which is $L = 2000$ m here. Because this value is the half-length of the fracture, which is on one side of the well, the total length is twice this value. Consequently, the fracture length is $L = 4000$ m.
- With reference to the geology data, the fracture height value is limited by the distance between the injection section and the cap formation. This value can be estimated on the basis of lithology data of the field where the injection well is located. In this example, this value is estimated as $H = 100$ m.
- Consequently, the volume of fluid with cuttings V can be calculated as:
- $V = W \times L \times H = 0.069 \times 4000 \times 100 = 2.76 \times 10^4 \, \text{m}^3$, with $1 \, \text{m}^3 = 6.29$ bbl.
- Usually, the volume of cuttings generated by drilling a well can be calculated by the bit size and total length of the well, considering the variation of bit sizes at different *TVD* intervals. The slurry concentration value of the injection fluid varies from approximately 10 to 20% because of the operational requirements related to the size of milled slurry. The details of this calculation are standard and are omitted here.

4.6 CONCLUSION

This work has established an integrated workflow for a feasibility study of cuttings reinjection (CRI) in the framework of 3D geomechanics. Various numerical solutions, such as stress distribution, fracture initiation and propagation, and cap integrity, as well as fault reactivation obtained with 3D finite element modeling, 2D finite element modeling, and 1D analytical modeling, have been used in the CRI decision-making process. As compared with the conventional CRI workflow, which is primarily based on an empirical method, this integrated workflow for a CRI feasibility study provides greater accuracy and higher efficiency.

A process for the 3D calculation of the IPW has been proposed. Finite element submodeling techniques were adopted for accuracy and efficiency of the hydraulic fracturing analysis. Two submodels were proposed for fracture initiation and propagation in horizontal and vertical directions, respectively. In this way, the computational burden caused by the fracturing analysis has been reduced significantly.

A process to estimate cap integrity has been proposed. The submodel established in Section 4.4.8.5 for fracturing analysis in the vertical direction has been used. Cap integrity has been validated by two methods:

1. Numerical solution fracture propagation has been checked. Cap integrity is ensured with the phenomena that the induced fracture stops at the bottom of the cap formation.
2. The value of injection pressure for initiating fracture at the cap information is checked by setting the injection point at the bottom of the cap formation. Cap integrity is ensured as long as the injection pressure required to generate the fracture at the cap formation is significantly higher than the value of injection pressure for fracture propagation at the stable propagation stage.

A process of fault reactivation analysis has been proposed. Fluid migration and seismic analysis have been performed in this process. A 2D plane-strain finite element model is used to provide accuracy and efficiency. A semi-analytical method is used for the calculation of seismicity magnitude by using the numerical solution of displacement discrepancy across the fault as input into the analytical equations.

A discussion of the fracture width value was presented for the purpose of connection results obtained with the simplified 3D geomechanical model with its application in the CRI practice. In general, if the perforation section is located in the lower part of the formation for CRI, this value of fracture width is acceptable to use for the calculation of the accommodation volume; if the location of the perforation section is in the upper part of the CRI target formation, the accommodation volume could be smaller than the value calculated.

CHAPTER 5

Geomechanics-based wellbore trajectory optimization for tight formation with natural fractures

The selection of an optimized wellbore trajectory is one of the key factors that determine the success of hydraulic fracturing. This chapter presents the geomechanics-based wellbore trajectory optimization.

Geomechanics-based wellbore trajectory optimization uses a geomechanics analysis solution, such as geostress components values. For calculation convenience, the F_p-potential concept is proposed as an index for selecting the optimized wellbore trajectory.

In the simplest case, the wellbore trajectory of a tight formation should be taken in the direction of maximum horizontal stress (SH) to maximize the stimulated reservoir volume; in this case, the injection-induced fracture will propagate in the direction that is perpendicular to the axis of wellbore trajectory.

When a natural fracture exists in the tight formation, injection-induced fractures are believed to be determined by those natural fractures that can open under hydraulic injection stimulation. Consequently, the task of trajectory optimization is to locate those natural fractures that are easy to open with injection. These fractures are also known as critically stressed fractures (CSF).

The workflow for selecting the optimized wellbore trajectory based on geomechanics solutions and the concept of CSF is presented in this chapter.

For trajectory optimization focused on the primary fracturing design, the workflow is performed with the initial geostress. For the optimization focused on the fracturing design for a field that is disturbed by primary fracturing from offset wells, the workflow is performed with the stress field disturbed by primary fracturing.

5.1 INTRODUCTION

Tight formation refers to shale gas/oil formations and tight sand gas/oil formations. Hydraulic fracturing is essential for production in tight formations. Trajectory optimization is a fundamental aspect of a wellbore design in a tight formation with natural fractures (Bond *et al.*, 2006; Himmerlberg and Eckert, 2013; Manchanda *et al.*, 2012). A deliberately optimized wellbore trajectory enables maximum efficiency of hydraulic fracturing; the value of stimulated reservoir volume (*SRV*) will reach its maximum if the trajectory is selected to go through the right place (sweet spot) and is in right direction.

CSF is a concept to represent a type of natural fracture in various tight formations. It defines those fractures being at stress status, which approaches the status of opening governed by Mohr-Coulomb type criteria (Franquet *et al.*, 2008). Critically stressed fractures are easy to open by hydraulic fracturing measures, and thus, the optimized trajectory designed to go through tight formation should go through the locations of CSFs in the formations.

The trajectory optimization of a wellbore in formations with natural fractures has another factor to be considered in addition to the previously described critical stress status. In general, when an angle exists between the direction of natural fractures and the direction of the *SH*, the propagation of the fracture generated by injection will occur in the direction that is determined by several factors, including the local directions of principal stresses and directions of natural fractures.

73

Section 5.2 presents the workflow and steps for determining the optimized trajectory in terms of geomechanics solutions and the CSF concept. Validation examples with initial geostress and geostructure data from an actual project are presented. Section 5.3 presents the workflow and trajectory optimization steps focused on a re-fracturing design. In this case, the selection of an optimized trajectory depends on additional factors, rather than on the CSF principle only.

5.2 DETERMINING OPTIMIZED TRAJECTORY IN TERMS OF THE CSF CONCEPT

The task of this work is to select an optimized well trajectory in terms of the CSF concept. Equation (5.1) defines the Mohr-Coulomb criterion for the occurrence of frictional slip along surfaces of the fracture. It is also used to judge the loading condition of a stress point of a finite element model:

$$\tau = c + \mu_f \sigma_n \tag{5.1}$$

where τ is the shear stress at the frictional surface; c is the cohesive strength of the formation; μ_f is the internal frictional coefficient and $\mu_f = \tan(\varphi)$, φ is the internal frictional angle; and σ_n is the normal stress at the plane of critical stress status. For a natural fracture at a location within the formation for which the relationship listed in Equation (5.1) is approximately satisfied, it is regarded as being at a critically stressed status. A fracture with a stress that would be classified as critically stressed is regarded as CSF.

5.2.1 **Workflow for the selection of an optimized trajectory**

The goal and task for optimizing the trajectory of a wellbore within a tight formation with natural fractures is to locate the region of these CSFs in terms of the initial geostress field. For this task, The CSF potential F_p concept is defined as:

$$F_p = \tau - (c + \mu_f \sigma_n) \leq 0 \tag{5.2}$$

For numerical calculation convenience with the finite element method (Owen and Hinton, 1980), Equation (5.2) is rewritten in the form of principal stresses function. Here we can write the CSF potential F_β on the surfaces of fractures with inclination angle β as:

$$F_\beta = \tau_\beta - (c_\beta + \mu_\beta \sigma_{n\beta}) \tag{5.3}$$

where subscript β indicates a parameter being defined on the fracture surface with inclination angle β.

The CSF potential F_β in Equation (5.2) will be used as a geomechanics index of a fracture to distinguish the fractures that are approximately at a critically stressed status from those fractures that are significantly NOT critically stressed. The values of F_β will be calculated at each integration point of a finite element model with the User Subroutine (Dassault Systèmes, 2010). The optimized trajectory of a wellbore is the trajectory in which the F_β point values on this trajectory are larger than the F point values along other trajectories:

$$F_{\text{optimized}} \geq F_{\text{other}} \tag{5.4}$$

Steps for optimizing wellbore trajectory in terms of CSF by 3D numerical method include the following:

1. Perform 1D geomechanics analysis with logging data and other measured data, such as leakoff tests from single wells.
2. Construct the 3D initial stress field with the given geostructure and solutions of the 1D geomechanics analysis.
3. Perform 3D finite element analysis for the solution of initial geostress field. Values of F_β will be calculated at each point of the mesh of the formations.

Figure 5.1   The finite element mesh of the model.

4. Select the optimized trajectory by values of F_β in terms of Equations (5.3) and (5.4) by checking and comparing values of F_β of each point.

5.2.2   Numerical application

The following model of the application is provided as an example to validate the workflow proposed in previous subsections.

5.2.2.1   *1D geomechanics solutions*

The vertical stress value of the initial stress field is generated by gravity, with an averaged density value from top to bottom as 2400 kg/m³.

The lateral stress factors are: 0.875 for maximum horizontal stress *SH* which is selected as σ_x, and 0.5 for minimum horizontal stress σ_y.

The pore pressure value is set as 85 MPa. The initial void ratio is 0.1.

The Young's modulus value is 20 GPa, and Poisson's ratio is set as 0.2. These parameter values are used for all formations in the model.

5.2.2.2   *Construct the 3D initial stress field*

As shown in Figure 5.1, the model includes three formation layers: Upper formation, Reservoir formation, and Bottom formation. The Reservoir formation is a tight sand gas formation; its thickness values vary from 120 to 170 m. Natural fractures are distributed throughout the reservoir formation. For simplicity, no dominant directions of natural fractures are considered. Consequently, the distribution of natural fractures is considered to be isotropic. The depth of the reservoir is approximately *TVD* = 5000 m.

Figure 5.2 shows the geometry and mesh of the reservoir formation.

For brevity, the model provided here includes only the reservoir formation and the formations directly above and below. The thickness of the model is approximately 850 m, which corresponds to a *TVD* depth interval from 4600 to 5450 m. Its width is 2400 m, and its length is 3700 m.

There is one set of natural fractures existing in the reservoir formation, and their azimuth angle is 90°, inclination angle is $\beta = 73°$. Details about definition of directional parameters of a set of natural fractures can be found in Chapter 8.

5.2.2.3   *3D Finite element analysis for the solution of the initial geostress field and F*

The boundary conditions of the model are zero normal displacement constraints to all lateral surfaces and to the top and bottom.

The overburden stress will be modeled by means of the vertical stress and traction generated by the boundary conditions on the top surface.

Figure 5.2   Mesh of the reservoir formation.

S, S33
(Avg: 75%)
−3.589 × 10^7
−4.289 × 10^7
−4.989 × 10^7
−5.690 × 10^7
−6.390 × 10^7
−7.090 × 10^7
−7.790 × 10^7
−8.491 × 10^7
−9.191 × 10^7
−9.891 × 10^7
−1.059 × 10^8
−1.129 × 10^8
−1.199 × 10^8

Figure 5.3   Contour of vertical stress.

Figures 5.3 through 5.6 show the numerical results of the contours of the vertical stress component, von Mises stress, pore pressure, and F within the reservoir.

Because of the existence of pore pressure in the reservoir formation, the effective stress values there are less than those of the other formations.

As shown in Figure 5.6, the values of the F_β-contour vary in the range of −20 to 38 MPa, which are greater than 0. The reason for this is that only the porous elastic constitutive model is used, and no plastic flow is performed in the calculation. This set of numerical results of F_β is only an elastic approximation of the actual case. In reality, there will be plastic flow at the material point when the status of $F_\beta = 0$ is reached. In this case, $F_\beta = 0$ is the plastic loading criterion.

5.2.2.4   *Selection of optimized trajectory*
The F_β-contours shown in Figure 5.6 and 5.7 indicate the locations of CSF and show where natural fractures are easy to open by hydraulic fracturing. As shown in Figure 5.6, the value of F_β on the top of the anti-cline reservoir formation is shown to be less than that of the area some distance from the top. The optimized trajectory should be selected as on the paths represented by two bold curves. The values of F_β on the boundary surface should be ignored because of boundary effect.

Figure 5.7 shows the case of F_β-contour when *SH* is σ_y. The F_β-contour values on the left are much greater than those on the right. Consequently, CSFs are located in the left side of the reservoir formation. The optimized trajectory selections are made as shown by the three bold curves.

5.3   TRAJECTORY OPTIMIZATION FOCUSING ON A FRACTURING DESIGN FOR A DISTURBED FIELD

For trajectory optimization that focuses on the fracturing design for a field that is disturbed by primary fracturing from offset wells, the CSF calculation is performed on the basis of a stress field that is disturbed by primary fracturing. In this case, the effect of natural fractures and

Figure 5.4   Contour of von Mises stress.

Figure 5.5   Contour of initial pore pressure (80 MPa).

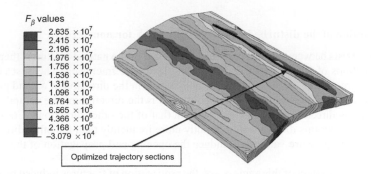

Figure 5.6   Contour of F_β within reservoir with *SH* as σ_x.

induced fractures generated by primary fracturing should be considered during the calculation of the disturbed stress.

A natural means of achieving an optimized trajectory solution with CSF for fracturing in disturbed field includes: (i) performing numerical modeling of the fracturing simulation, and (ii) using the stress field taken directly from the numerical results of the fracturing analysis. Chapter 8 shows the detailed fracturing simulation analyses for various situations. For this chapter, however, the focus is only on the analysis of using the disturbed fracturing stress solution to perform the CSF-based trajectory optimization.

With reference to the fracturing simulation provided in Chapter 8, two cases should be considered: (i) the principal directions of natural fractures are not the same as that of the initial geostress (i.e., an angle α exists between the direction of *SH* and that of the natural fractures), and (ii) the principal directions of the natural fractures overlap with that of the initial geostress.

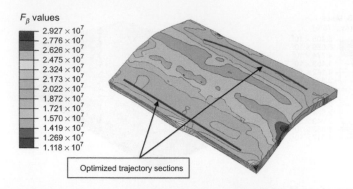

Figure 5.7   Contour of F_β-within reservoir with *SH* as σ_y.

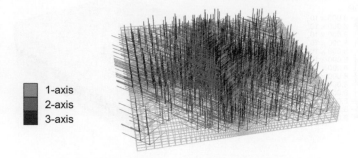

Figure 5.8   Mesh of the model and its directions of orthotropic permeability tensor.

5.3.1   The solution of the disturbed geostress field and F for non-zero α_{sf}

When angle α_{sf} exists between the direction of *SH* and that of the natural fractures, there are two competing directions of fracture propagation during hydraulic fracturing. The first competing direction is the direction of *SH*. The fracture usually opens in the direction of *Sh* and propagates in the direction of *SH*. The second competing direction is the direction of the natural fractures; permeability in this direction is usually greater than that of the other directions. Consequently, the porous flow follows this direction significantly more frequently than the other directions. As a result, greater pore pressure values will induce the opening and propagation of the fracture in this direction.

For brevity, only the effect of this angle α_{sf} on the propagation of fractures induced by hydraulic injection will be investigated with a simplified uniform stress field model, as shown in Figure 5.8.

The model shown in Figure 5.8 is 200 m in width and length, and 25 m in height. The 20 m thick reservoir is covered with a 5 m overburden layer. The principal directions of orthotropic permeability are given as: axis-1 is NE45°, axis-2 is NE135°, and axis-3 is vertical. Axis-1 is assumed to be the direction of natural fractures; consequently, it is the direction in which the permeability is greater than that of other two directions. Inclination angle of natural fractures is set as $\beta = 90°$.

The F_β potential value is the same throughout the model at the initial status before hydraulic injection because the initial stress field is uniform and the formation top is flat.

In this case, at the initial status, there is no direction along which an optimized trajectory can be selected. Hydraulic injection, however, will cause stress changes in the stimulated reservoir volume, and the transient contour of F_β potential can be determined during the hydraulic fracturing of a tight formation. Consequently, an optimized wellbore trajectory can be obtained in terms of these transient *F* contours.

Figure 5.9   Contour of F_β within reservoir at time $t = 17.53$ minutes.

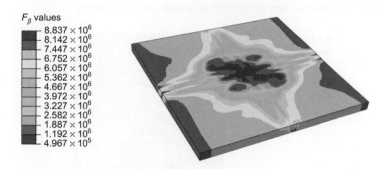

Figure 5.10   Contour of F_β within reservoir at time $t = 26.64$ minutes.

Chapter 8 provides the details of the model and introduces techniques for modeling the hydraulic fracturing of naturally fractured formations with continuum damage mechanics. The values of the initial geostress, initial pore pressure, and material parameters are all taken from those described in Equations (8.19) to (8.22). The maximum horizontal principal stress *SH* is in the direction of the *Y*-axis. Injection loading values come from the curve that is shown in Figure 8.5 of Chapter 8. The damage-dependent permeability model is used. Information shown in Figures 8.3 and 8.4 are also used in this calculation. The values of the model parameters are assumed for the purpose of an efficient workflow illustration.

Figures 5.9 through 5.12 provide the numerical solution of the transient contour of F_β potential during hydraulic injection at various time moments. As shown in Figure 5.9, the values of F_β potential in the area of greater pore pressure caused by injection are significantly greater than those of other locations at time $t = 17.53$ minutes. F_β-contour distributes in a rectangular pattern with the shorter side in the direction of NE45°.

Figure 5.10 shows the F_p-contour at time $t = 26.64$ minutes; at this time, the F-contour around the injection point distributes in a rectangular pattern in which the two sides are almost equal, and the slightly shorter side is in the direction of NE45°. In addition, the F_β-contour at this moment has fingers extending in four directions.

Figure 5.11 shows the pore pressure contour at time $t = 26.64$ minutes. The pore pressure contour distributes in a rectangular pattern with the shorter side in the direction of NE45°.

Figure 5.12 shows the von Mises stress contour, which provides an index of the geostress field. At this point, the pore pressure is restored to its initial value of 20.66 MPa. This geostress field status is the disturbed stress field, which has changed from its original uniform status by hydraulic fracturing. Figure 5.12 shows that the stress field is no longer uniform because of the stiffness variation of the formation caused by hydraulic fracturing.

Figure 5.13 shows the damage intensity contour for the case of $\alpha_{sf} > 0$. The localization of the damage value for this case is rather weak and does not form a band. This result complicates the

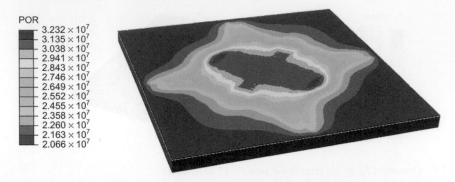

Figure 5.11   Contour of pore pressure within reservoir at time $t = 26.64$ minutes.

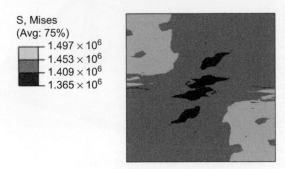

Figure 5.12   von Mises stress contour within the reservoir when the pore pressure is restored to its initial value.

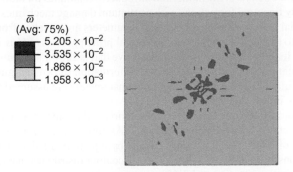

Figure 5.13   Contour of damage intensity within reservoir after fracturing, $\alpha > 0$.

optimized trajectory choice. Figure 5.14 shows the contour of F_β for CSF on the basis of stress field shown in Figure 5.12. Figure 5.14 shows the suggested optimized trajectories as curves AB and CD. These two curves have been selected to go through locations with the maximum F values in their neighborhood.

5.3.2   The solution of the disturbed geostress field and F for zero α_{sf}

Omitting the modeling details of fracturing (which are available in Chapter 8), this section provides the numerical results of the disturbed stress contours, damage intensity, and F-values for the case of $\alpha_{sf} = 0$.

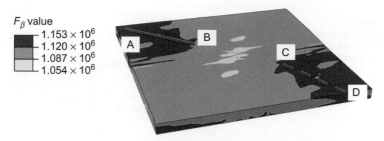

Figure 5.14   Contour of F_β within reservoir after fracturing and pore pressure unloading to initial reservoir pore pressure level, $\alpha_{sf} > 0$.

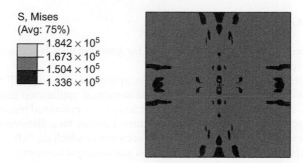

Figure 5.15   von Mises stress contour within the reservoir when the pore pressure is restored to its initial value.

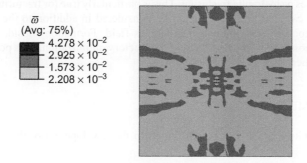

Figure 5.16   Contour of the damage intensity within the reservoir after fracturing, $\alpha = 0$.

As shown in Figure 5.15, the stress distribution has changed from its original uniform status to the current non-uniform status.

Figure 5.16 shows the distribution of the damage intensity caused by hydraulic fracturing. The localization of the damage distribution is rather strong, and several localization bands are formed. This result makes the optimized trajectory choice relatively easier.

Figure 5.17 shows the distribution of F_β values in terms of the geostress shown in Figure 5.15.

The suggested optimized trajectories are shown in Figure 5.17 as the two curves of AB and CD. These two curves penetrate the locations of maximum F_β values.

A comparison of Figures 5.14 and 5.17 indicates that (i) for the $\alpha_{sf} = 45°$ scenario, the F_β-contour is not the obvious path to select as the optimized trajectory; it is probably necessary to investigate the reason behind this phenomenon, and (ii) for the case of $\alpha_{sf} = 0$, the F_β-contour is the obvious path to select as the optimized trajectory.

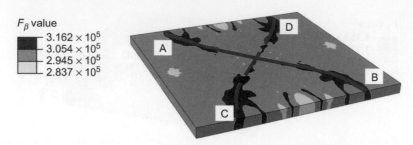

Figure 5.17   Contour of F_β within the reservoir after fracturing and pore pressure unloading to the initial reservoir pore pressure level, $\alpha_{sf} = 0$.

5.4   CONCLUDING REMARKS

This chapter introduces the principle and workflow for geomechanics-based wellbore trajectory optimization with a focus on hydraulic fracturing.

With the F_β-potential defined in Equation (5.3), the optimized trajectory will cross those areas with CSF in a given field of a tight formation. The numerical application indicates that a 3D F_β-contour provides an effective means for the selection of this optimized trajectory.

For a trajectory optimization focusing on the fracturing design for a disturbed field, two cases are considered: a scenario in which $\alpha_{sf} = 45°$ and a scenario in which $\alpha_{sf} = 0$.

When an angle exists between the natural fractures and principal horizontal stress component, it is not easy to select the optimized trajectory in terms of F_β-values. It is easier, however, to select the optimized trajectory for the scenario of $\alpha_{sf} = 0$.

Sometimes, the geomechanics factor F_β is not the only factor to consider for an optimized trajectory that focuses on hydraulic fracturing. This is particularly true for fracturing in a disturbed stress field. In this case, other factors should be considered in addition to the F_β-contour, as well as the time-dependent distribution of the stress field. Formation subsidence induced by production should possibly also be considered for trajectory optimization. This possibility should be investigated further.

ACKNOWLEDGEMENTS

The authors thank Guoyang Shen for contribution to the development of the User Subroutine used in this work.

CHAPTER 6

Numerical solution of widened mud weight window for drilling through naturally fractured reservoirs

This chapter introduces the concepts of the widened mud weight window (MWW) and cohesive cracks and presents a workflow that models and calculates a widened MWW for a wellbore within a naturally fractured formation. Hydraulic plugging and stress caging are numerically simulated and analyzed. Workflow and numerical models are presented to obtain a widened MWW through hydraulic plugging and stress caging. A validation example from a vertical well section within a subsalt formation in the Gulf of Mexico (GOM) is presented. Another application of the proposed workflow presented includes a case of a horizontal well that was drilled through a natural fractured shale gas formation.

6.1 INTRODUCTION

In many cases, the MWW for subsalt well sections is narrow. Natural fractures are a major contributor to mud weight loss during the drilling of subsalt well sections. Hydraulic fracturing and stress caging are popular operational measures to widen MWW in these cases, and various works have been presented (Edwards *et al.*, 2002; Guo *et al.*, 2014; Lietard *et al.*, 1996; Mehrabian *et al.*, 2015). The stress caging principle is used to fill and seal the natural fractures with lost circulation materials (LCMs).

Lietard *et al.* (1996) determined the width of a natural fracture using mud loss observations during drilling operations and further proposed a measure to prepare LCM for this set of natural fractures when the width under a given hydraulic pressure is known. Image logging data obtained using logging while drilling (LWD) tools was also discussed. Caughron *et al.* (2002) and Edwards *et al.* (2002) presented their successful practices using real-time geomechanics information to determine the fracture location and the optimized volume of LCM. Sanad *et al.* (2004) presented a wellbore enhancement study using a numerical method. A finite element analysis (FEA) model was used to show how the fracture re-initiation pressure or near-wellbore fracture gradient (*FG*) was increased by the *FG* enhancement squeeze-system treatment. Alberty and McLean (2004) presented a model that predicts early pressure buildup behavior and discussed how the model can improve the interpretation of formation-pressure integrity tests. Kageson-Loe *et al.* (2009) investigated the interplay between LCM, fluid loss, and formation permeability in the plugging and sealing of fractures. This investigation reports that the LCM blend that contained larger particles than the fracture aperture is necessary to form the most competent fracture sealing at the entrance to the aperture. Fett *et al.* (2009) reported a case of well strengthening in a depleted and highly unconsolidated sand formation in deepwater GOM.

Salehi and Nygaard (2011, 2012) investigated the hypothesis that wellbore hoop stress increases when fractures are wedged and/or sealed during lost circulation control operations. Salehi and Nygaard (2011, 2012) came to a rather controversial conclusion: fracture wedging cannot increase wellbore hoop stress more than its ideal state, wherein no fracture exists. However, it will help to restore part or all of the wellbore hoop stress lost during fracture propagation, which can act as a secondary mechanism to increase wellbore *FG*.

Savari *et al.* (2014) presented an approach for designing LCMs to withstand higher wellbore pressures within a fracture. By using a permeability plugging apparatus (PPA) with tapered slots,

83

a plug breaking pressure (PBP) property was determined for various LCM combinations. Al-saba *et al.* (2014) presented an evaluation of sealing wide fractures (e.g., 2 mm wide) using conventional LCM. In their work, LCM sealing efficiency was defined as the seal/bridge maximum breakdown pressure. Tests were conducted on a fit-for-purpose apparatus designed to evaluate LCM performance by measuring the sealing efficiency under high pressures and temperatures. Mehrabian *et al.* (2015) presented a study on the geomechanics of lost circulation events and wellbore-strengthening operations. The study explored these wellbore-strengthening mechanisms through an analytical solution to the related solid-mechanics model of the wellbore and its adjacent fractures. It also investigated the wellbore hoop stress enhancement upon fracturing.

These referenced works indicate that a widened MWW can be obtained using hydraulic-plugging and stress-caging measures.

During hydraulic-plugging operations, LCM particles are pushed into fractures within the formation around the wellbore under pressure, which can open those fractures. Consequently, the fracture permeability is reduced, and mud loss is avoided for a certain pressure value range.

The stress-caging concept theorizes that hoop stress in the formation near the wellbore surface is increased by filling the LCM particles into natural and induced fractures within the formation around the wellbore. Thus, it effectively increases the *FG* of the formation; fracture opening requires that the hoop stress be first overcome. Therefore, the increased hoop stress forms a cage of enhancement in the formation near the wellbore surface. However, some research presents evidence that a stress cage cannot be formed in the reported cases (Salehi and Nygaard, 2011, 2012).

There are sufficient experimental results regarding hydraulic plugging and related operational issues reported in these referenced studies. There is little information about the description of natural fractures and their effect on the practice of widening MWW, although the permeability and mechanical properties, such as stiffness of the fracture and fracture propagation resistance, are crucial to the effectiveness of the wellbore-strengthening measures.

The elastic modulus stress dependency phenomenon, as well as strength parameters, are essential mechanical properties for an unconsolidated sand formation (Shen *et al.*, 2010). With the greater stress value, formation compaction causes hardening of the unconsolidated sand formation. This is an important factor that can affect the stress cage formation around the wellbore in unconsolidated formations.

This chapter investigates the behavior of natural fractures under injection load using a three-dimensional (3D) FEA method. The following contents are modeled in detail:

- Fracture opening and propagation under injection into a natural fracture.
- Stress cage: variation of hoop stress around the wellbore related to fracture opening and propagation; hoop stress values during and after the fracturing process validated.
- Initial width of natural fracture.
- Natural fracture mechanical properties and their effect on the stress cage.
- Effect of natural fractures permeability on the effective strengthening; an illustrative example is presented at the end of the chapter.
- A case of stress cage in a wellbore from deepwater GOM.

6.2   MODEL DESCRIPTION: THEORY

6.2.1   **Constitutive model**

The model used in this chapter is the quasi-brittle fracture (Bazant and Cedolin, 1991; Cocchetti *et al.*, 2002; Hillorberg *et al.*, 1976). Figure 6.1 shows a quasi-brittle fracture in the sandstone formation; a damage zone exists ahead of the macroscopic fracture. The continuum damage concept (Lee and Fenves, 1998; Lubliner *et al.*, 1998) is used to describe the initiation and growth of microcracks in the damage zone. The material is not yet broken completely, and tensile strength

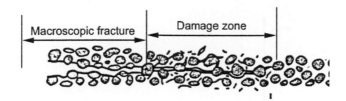

Figure 6.1   Illustration of a crack in the rock formation.

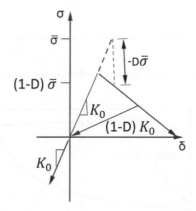

Figure 6.2   Illustration of typical damaged response.

or cohesive strength still exists within the damage zone. Fracture propagation is modeled with damage evolution in the damage zone ahead of the fracture.

Figure 6.2 shows the damage variable D and its relationship with stiffness K and effective stress $\bar{\sigma}$ and fracture opening δ. Equation 6.1 shows the relationship between the effective stress $\bar{\sigma}$ and damage variable D:

$$\sigma = (1 - D)\bar{\sigma} \tag{6.1}$$

where D is the scalar damage variable, and $D = 0$ for undamaged intact material; $D = 1$ fully damaged; D monotonically increases.

6.2.2   Damage initiation criterion

Traction separation law is used. This law is defined by peak strength N and fracture energy G_{TC} as expressed in Equation (6.2):

$$\text{Max} \left\{ \frac{\langle \sigma_n \rangle}{N_{max}}, \frac{\sigma_t}{T_{max}} \frac{\sigma_s}{S_{max}} \right\} = 1, \ \langle \sigma_n \rangle = \sigma_n \ \text{for} \ \sigma_n > 0; \ \text{and} \ = 0 \ \text{for} \ \sigma_n < 0 \tag{6.2}$$

where σ_n is the stress in the normal direction to the fracture surface, and σ_t and σ_s are stress components in the two tangential directions. N_{max}, T_{max}, and S_{max} are strength of material in normal direction and two tangential directions, respectively.

6.2.3   Damage evolution law

The damage evolution law used in this chapter is based on energy. The Benzeggagh-Kenane (BK) law expressed in Equation (6.3) is used to define the damage evolution:

$$G_{IC} + (G_{IIC} - G_{IC}) \frac{G_{shear}}{G_T} = G_{TC} \tag{6.3}$$

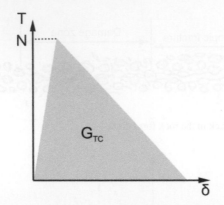

Figure 6.3   Illustration of traction-separation response on the fracture surface.

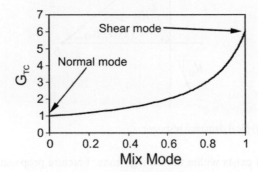

Figure 6.4   Illustration of dependence of fracture energy on mix mode.

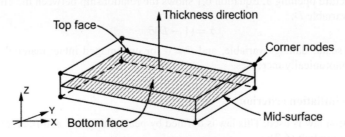

Figure 6.5   Thickness direction of the cohesive element Coh3D8P.

where $G_{\text{shear}} = G_{\text{II}} + G_{\text{III}}$, and $G_{\text{T}} = G_{\text{I}} + G_{\text{shear}}$. Figure 6.3 shows the relationship between fracture energy G_{TC} and the maximum tensile strength N of the material. Figure 6.4 shows the variation in value of fracture energy with mix mode of fracture.

6.2.4   Finite element type: the cohesive element

Figure 6.5 shows the Coh3D8P element is the cohesive element coupled with pore pressure (*PP*) designed to simulate rock fracturing under hydraulic fracture (Dassault Systèmes, 2011). Figure 6.6 shows the number sequence of the cohesive element Coh3D8P. Although it is labeled as Coh3D8P, this type of element actually comprises 12 nodes. The nodes on the midsurface are specially designed for fluid leakoff calculations.

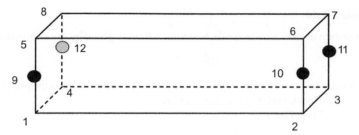

Figure 6.6   Number sequence of the cohesive element Coh3D8P. Midsurface nodes are specially designed for fluid leakoff calculations.

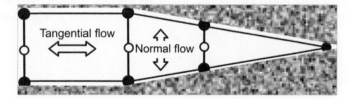

Figure 6.7   Flow within cohesive elements.

The stress components designed for this type of cohesive element include the following:

- S33: direct through-thickness stress.
- S13: transverse shear stress.
- S23: transverse shear stress.

The tensors of the mechanical variables of the same type of cohesive element, such as strain, all have the same number of components. The displacement and *PP* are given as nodal variables in the calculation of this type of cohesive element.

6.3   FLUID FLOW MODEL OF THE COHESIVE ELEMENT

In the applications of the cohesive element in hydraulic fracturing operations, the fluid flow continuity within the gap and through the interface should be maintained to enable the fluid pressure on the cohesive element surface to contribute to its mechanical behavior. This enables the modeling of the hydraulically driven fracture. It also enables the modeling of an additional resistance layer on the surface of the cohesive element.

6.3.1   Defining pore fluid flow properties

The fluid-constitutive response comprises two parts: tangential flow within the gap, which can be modeled with either a Newtonian or power-law model, and normal flow across the gap, which can reflect resistance as a result of caking or fouling effects. Figure 6.7 shows the flow patterns of the pore fluid in the element.

6.3.2   Tangential flow

By default, there is no tangential flow of pore fluid within the cohesive element. To enable a tangential flow, a gap flow property should be defined in conjunction with the pore fluid material definition in the model input data. There are two types of tangential flow that can be adopted in the calculation: Newtonian fluid and power-law fluid. Their expressions are given in the following subsections.

6.3.3 **Newtonian fluid**

The volume flow-rate density vector $\mathbf{q}$ of a Newtonian fluid is given by the expression:

$$\mathbf{q}Z = -k_t \nabla p \tag{6.4}$$

where k_t is the tangential permeability (the resistance to the fluid flow), ∇p is the pressure gradient along the cohesive element, and Z is the gap opening.

The gap opening Z is defined as:

$$Z = t_{curr} - t_{orig} + g_{init} \tag{6.5}$$

where t_{curr} and t_{orig} are the current and original cohesive element geometrical thicknesses, respectively, and g_{int} is the initial gap opening, which has a default value of 0.002 m.

With reference to the Reynolds equation, the tangential permeability k_t is given as:

$$k_t = \frac{z^3}{12\mu} \tag{6.6}$$

where μ is the fluid viscosity.

6.3.4 **Power-law fluid**

For a power-law fluid, its constitutive relation is defined as:

$$\tau = K_f \dot{\gamma}^\alpha \tag{6.7}$$

where τ is the shear stress, $\dot{\gamma}$ is the shear-strain rate, K_f is the fluid consistency, and α is the power-law coefficient. Its tangential volume flow-rate density is defined as:

$$\mathbf{q}Z = -\left(\frac{2\alpha}{1+2\alpha}\right)\left(\frac{1}{K}\right)^{\frac{1}{\alpha}}\left(\frac{d}{2}\right)^{\frac{1+2\alpha}{\alpha}}\|\nabla p\|^{\frac{1-\alpha}{\alpha}}\nabla p \tag{6.8}$$

6.3.5 **Normal flow across gap surfaces**

Normal flow is defined through a fluid leakoff coefficient for the pore fluid material. This coefficient defines a pressure-flow relationship between the middle nodes of the cohesive element and their adjacent surface nodes. The fluid leakoff coefficients can be interpreted as the permeability of a finite layer of material on the cohesive element surfaces.

With the leakoff coefficients c_t and c_b, the normal flow is defined as:

$$q_t = c_t(p_i - p_t) \tag{6.9}$$
$$q_b = c_b(p_i - p_b) \tag{6.10}$$

where q_t and q_b are the flow rates into the top and bottom surfaces, respectively, p_i is the midface pressure, and p_t and p_b are the *PP*s on the top and bottom surfaces, respectively.

The following output variables are available when flow is enabled in *PP* cohesive elements:

- Gap fluid volume rate (*GFVR*).
- Fracture opening (*PFOPEN*).

6.4 VALIDATION EXAMPLE: WIDENED MUD WEIGHT WINDOW FOR SIMPLE CASES

This example simulates fracture opening within the target formation at a true vertical depth (*TVD*) of 1000 m.

Figure 6.8   Geometry of the model.

The model geometry is simplified in this example. Only a slide of a semicircular plate in a horizontal plane is adopted, and fracture propagating in the vertical direction is simplified. The focus of the analysis is on the fracture propagation in the horizontal direction.

6.4.1   Geometry

Figure 6.8 shows a model of a semicircular layer that is adopted for the geometry of the formations investigated. The layer is 0.121 m (5 in.) thick with a 0.242 m (10 in.) borehole in the center. The radius of the model is 3 m.

The locus of the natural fracture, which will be opened by hydraulic injection pressure, is set at the central surface of the semicircular plate. A set of cohesive elements that have both PP and displacement as their primary nodal variables are used to simulate the fracturing behavior.

6.4.2   Initial conditions

The initial conditions for the model are defined as follows:

- Void ratio is 0.35 for the reservoir formation.
- Initial PP is 1.0 MPa for all formations; simplifications were made for this initial condition.
- Initial geostress components of all formations are given the values in effective stress space as $\sigma_x = -10$ MPa, $\sigma_y = -15$ MPa, $\sigma_z = -20$ MPa. All shear-stress components are zero. The total overburden stress (Sv) is –21 MPa, which is the sum of the vertical effective stress and the PP.

A value of 0.5 mm as the initial gap is assigned to the nodes at which the crack begins under the given hydraulic pressure.

6.4.3   Boundary condition

Displacement constraints were applied to the normal directions of the outer surface, symmetry surface, and bottom and top surfaces. The pore pressure boundary conditions of the symmetry surface are given as zero. For other boundary surfaces, the pore pressure values are assumed to be constant and are the same values as those used for the initial PP.

6.4.4   Loads

The loads applied in this model include the following:

- Geostress and gravity load.
- Concentrated flow during the injection step.
- Mud-weight pressure applied to the inner surface of the wellbore.

Figure 6.9a shows the location of the injection flow application at given nodes (red dots). Meanwhile, uniform surface pressure, which equals the value of injection pressure and varies with time, is applied to the wellbore surface (Fig. 6.9b). This is used for the numerical simulation

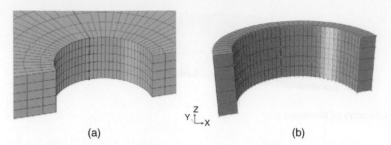

(a)   (b)

Figure 6.9   (a) Nodes marked as red dots represent the locations at which injection flow loadings are applied; (b) uniform surface pressure applied to the wellbore surface.

Table 6.1   Values of material parameters for cohesive element.

Damage initiation, criterion = *MAXS*	S1 [Pa]	S2 [Pa]	S3 [Pa]
	1.0	10.0	10.0
Damage evolution, type = *ENERGY*	G1 [N/m]	G2 [N/m]	G3 [N/m]
	0.1	10.0	10.0
Density [kg/m^3]		2000	
Elastic parameters	Kn [Pa]	Kt1 [Pa]	Kt2 [Pa]
	8.5×10^{10}	8.5×10^{10}	8.5×10^{10}
Fluid leakoff and gap flow	k-bottom [m/s]	k-top [m/s]	K [m/s]
	1×10^{-12}	1×10^{-12}	1×10^{-8}

Table 6.2   Values of material parameters for formation matrix.

Elastic parameters	E [Pa]	v
	9×10^{-9}	0.27
Permeability/hydraulic conductivity [m/s]	5×10^{-12}	
Void ratio	0.35	
Density [kg m^{-3}]	2100	

of the fracture opening and propagation. This surface pressure load is a constant value and no longer varies with time when numerical modeling is used to estimate the efficiency of stress cage *t*. This is performed because for that moment, injection pressure is removed and only the normal static-pressure loads remain on the wellbore surface. Therefore, it is a constant that is related to the mud-weight density only.

6.4.5   Values of material parameter

Tables 6.1 and 6.2 list the values of material parameters used in the calculation.

In Table 6.1, S1 is the strength value in the normal direction of the fracture surface. S2 and S3 are strength values in the two directions tangential to the fracture surface. G1, G2, and G3 are the values of fracture energy for each single-mode fracture in the directions normal and tangential to the fracture surface, respectively. Kn, Kt1, and Kt2 are interfacial stiffness in the directions normal and tangential to the fracture surface, respectively.

In Table 6.2, E is Young's modulus, and v is Poisson's ratio. Elastic behavior is assumed for the reservoir properties.

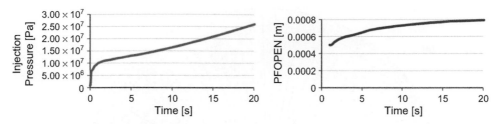

Figure 6.10   (a) Variation of injection pressure *P* with time *t* at a constant injection flow rate; (b) variation of fracture opening with time-dependent injection pressure shown in (a).

6.4.6   Procedure for numerical simulation of natural fracture opening under injection

The procedure for numerical simulation of fracture opening under injection using 3D FEA method includes the following steps:

- Build an initial geostress field, and set the initial values of the natural fracture initial gap.
- Simulate the drilling of the wellbore with the mud-weight pressure applied on the wellbore surface. This is modeled by removing elements of the wellbore and simultaneously applying pressure load to the wellbore surface.
- Perform the first step of the consolidation analysis, which simulates fluid injection into the formation under pressure. In this process, the cohesive element that represents the natural fracture will open as a result of the fracturing.

6.4.7   Numerical results Case 1: injecting process, fracture opening, and propagation

Figures 6.10 through 6.12 present the numerical results of injection pressure variation with time, fracture opening, and contour of fracture propagation.

Figure 6.10a shows that injection pressure increases with time during this loading period. Because of the values of permeability assigned to the formation, it does not decrease after its peak value in the middle of the loading process. Figure 6.10b shows the variation of fracture opening with time. The maximum value of fracture opening is 0.00077 m, which is 0.77 mm for a natural fracture with and original width of 0.5 mm. This maximum value is approximately achieved with injection pressure 20 MPa at time $t = 15$ seconds in this model.

Figure 6.11a shows that the fracture mouth opens with the increase of injection pressure. However, Figure 6.11b shows that the mouth is closed when the fracture opening propagates into its second stage in which the fracture opening moves away from the wellbore. Obviously, this secondary stage is not good for wellbore strengthening, thus should be avoided in practice. It should be noted that Figure 6.11a is a zoomed view which only shows inner part of the model geometry, and Figure 6.11b shows the whole view of the model geometry.

Figure 6.12a shows variations of hoop stress around the wellbore along Path 1 in Figure 6.12b during injection. The time moment taken for illustration is $t = 0.2$, 10, 18, and 20 s.

6.4.8   Numerical results Case 2: static process after injection, fracture remains open

Figures 6.13 and 6.14 show the numerical results of static process after injection, fracture remains open. This process is modeled by a constant surface pressure along with a set of fracture-opening values. This set of fracture-opening values is realized by the related injection pressure value. In short, the injection loading is the same as Case 1, but only the surface pressure remains constant. Thereby, injection pressure and surface pressure are two independent factors. Injection flow is used to obtain the fracture opening value, and surface pressure is used to model static mud-weight pressure.

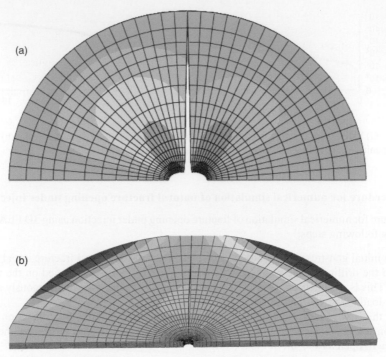

Figure 6.11   (a) Geometry of fracture opening in the primary stage of injection; (b) geometry of fracture in the secondary stage of propagation.

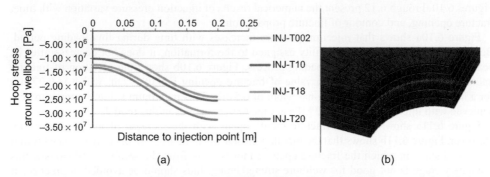

Figure 6.12   (a) Distribution of hoop stress around the wellbore at various time moment t along Path 1; (b) location of points of observation in Path 1 (red-dots on the line).

Figure 6.13a shows that the variation of injection pressure with time for Case 2 is similar to that shown in Figure 6.10a for Case 1. Figure 6.13b shows that the maximum fracture opening value for Case 2 (S-PFOPEN) is significantly smaller than in Case 1 (INJ-PFOPEN). The maximum fracture opening value for Case 1 is 0.8 mm. However, for Case 2, it is 0.7 mm for the initial fracture of 0.5-mm width, which represents a 10% difference.

Figure 6.14 shows the comparison of hoop-stress variation around the wellbore with time t for both Case 1 and Case 2. Time moments taken for illustration are $t = 0.2$, 10, 18, and 20 s. In Figure 6.14, "INJ" represents the curves of Case 1 during the injection process, and S represents the static pressure stage in Case 2.

Figure 6.14 shows that the values of hoop stress at time $t = 0.2$ seconds are the same in both cases. The value of hoop stress distribution around the wellbore increases with time as the injection

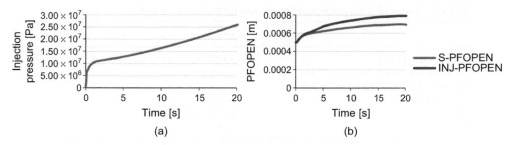

Figure 6.13   Case 2: (a) variation of injection pressure P with time t at a constant injection flow rate; (b) variation of fracture opening of Case 2 and its comparison with Case 1.

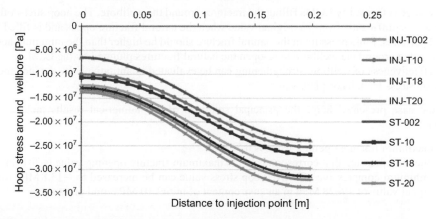

Figure 6.14   Hoop-stress distribution around the wellbore at various time moment t along Path 1 as shown in Figure 6.12b.

pressure increases, with peak value at $t = 20$ seconds. The difference between the two hoop stress curves at the same time moment $t = 10$ seconds is significant. The difference in Case 2 is greater than that of Case 1. This difference remains during the fracture opening and propagation process.

The hoop stress value at $t = 0.2$ s can be used as the original value for a wellbore with a natural fracture under initial static pressure. Figure 6.14 shows that the hoop stress values around the wellbore significantly increase from the previously described original. Consequently, it can be concluded the stress cage is effectively formed.

6.5   REMARKS

The following list provides a summary:

- Hydraulic plugging is the process that injection pressure applies to both the fracture and the wellbore surface. This surface pressure increases with time, whereas the injection process continues with a constant injection flow rate.
- The stress cage is formed when LCM is pushed/squeezed into the natural fractures, injection pressure disappears, and only static mud-weight pressure exists on the wellbore surface.

The following conclusions can be derived from the numerical results presented:

- Hoop stress increases with increments of injection pressure and fracture opening during hydraulic plugging with LCM squeezed into fractures. At the same time, the increased surface pressure applied on the wellbore surface reduces the value of compressive hoop stress by expanding the circumferential length of the wellbore surface.

Table 6.3   Values of stress and material parameters for the formation matrix.

	Total stress	Effective stress
Vertical stress [MPa]	−198.93	−68.98
Minimum horizontal stress (Sh) [MPa]	−155.05	−25.10
Maximum horizontal stress (SH) [MPa]	−155.74	−25.79
PP [MPa]	−129.95	
Young's modulus [MPa]	1.975×10^4	
Poisson's ratio	0.19	

- Stress cage is formed by LCMs filling the fractures around the wellbore. The hoop stress value increases significantly from its original value when the natural fracture opens and is filled by LCM particles. Fluid pressure in the natural fracture should be higher than the minimum hoop stress value around the wellbore to re-open the natural fractures plugged by the LCM.
- The effectiveness of wellbore strengthening can be evaluated quantitatively by the increment of hoop stress from its original value.

For the case presented here, the principal solutions of the numerical results include the following:

- The maximum fracture opening value is 0.77 mm.
- The injection pressure that corresponds to the maximum fracture opening value is 20 MPa.
- The minimum compressive effective hoop stress value can be increased to −13.8 MPa from its original value of −6.28 MPa in an environment of initial 10 MPa mud pressure.

6.6   CASE STUDY 1: WIDENED MUD WEIGHT WINDOW (MWW) FOR SUBSALT WELL IN DEEPWATER GULF OF MEXICO

The case study of a subsalt well in deepwater GOM includes the following information: its target formation location is at $TVD = 9616$ m (31,550 ft). Its pore pressure gradient is 132.2 MPa (11.5 ppg), and the planned mud weight for this section is 144.2 MPa (12.5 ppg). The initial width of the natural fracture is 0.5 mm. The objective is to calculate the maximum injection pressure to strengthen the wellbore with a stress cage being formed. Stress-cage estimation using various fracture opening values is necessary to prepare the LCM particle sizes.

Table 6.3 lists the aforementioned data transformed into a metric system. The data set listed in Table 6.1 is adopted for the other parameter values.

Figure 6.15 illustrates the workflow used in this work. Its principle has been described in previous sections.

6.6.1   **Numerical results**

Figures 6.16 and 6.17 show the numerical solutions of the fracture opening and the variation of injection pressure with time that corresponds to this set of fracture opening values. Figures 6.16a and 6.17a illustrate the overview of the time period, and Figures 6.16b and 6.17b zoom in on the variable values that are more practical.

Figure 6.18 shows the hoop stress distribution around the wellbore at various time moment t along Path 1, which is the same form shown in Figure 6.12b. Time moment is defined as $t1 = 1.37$ s, $t2 = 1.57$ s, $t3 = 1.77$ s, $t4 = 2.17$ s, and $t5 = 2.37$ s. Its PFOPEN values that correspond to the stress-cage stage are 0.500005, 0.500006, 0.528634, 0.610719, and 0.663842 mm. With the fracture-opening value for time $t5$, the value of minimum compressive hoop stress increases to

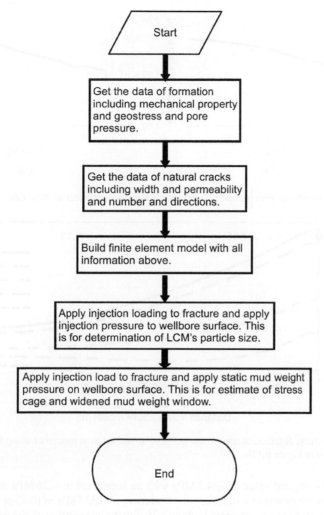

Figure 6.15   Numerical scheme flowchart to calculate widened MWW.

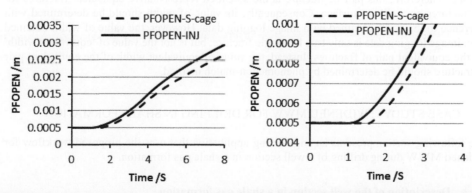

Figure 6.16   Fracture-opening variation caused by injection and its comparison with the caging-stage variation.

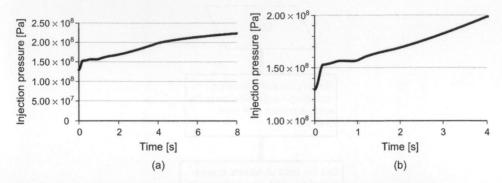

Figure 6.17   Variation of injection pressure P with time t at a constant injection flow rate.

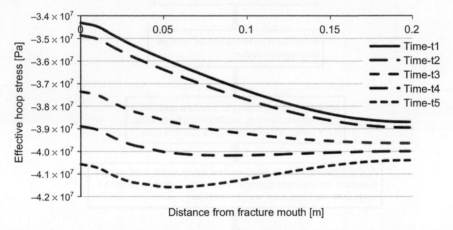

Figure 6.18   Hoop-stress distribution around the wellbore at various time moment t along Path 1, which is the same form shown in Figure 6.12b.

−40.6 MPa from its original value of −34.3 MPa with an increment of 6.26 MPa of compressive hoop stress. This corresponds to a stress gradient increase of 0.537 kPa/m (0.55 ppg).

The 0.537 kPa/m (0.55 ppg) increment is obtained using the assumption of one pair of fractures around the wellbore only. This pair of fractures represents all the fractures existing on wellbore surface. Therefore, this pair of fracture is the so-called representative equivalent fractures for all fractures on wellbore surface. Consequently, its value of width should be determined with reference to the width obtained with image logging data. However, the value of width obtained from image logging data is only true for single fracture, but is not the value of 'equivalent width' for the equivalent pair of fractures. Instead, the value of equivalent width of the equivalent pair of fracture should be determined by phenomenon-match method.

6.7   CASE STUDY 2: WIDENED MWW FOR DRILLING IN SHALE FORMATION

The following section presents an engineering application that uses the proposed workflow for widened MWW during drilling of a well section in a shale gas formation.

6.7.1   Description of the well section in a shale gas formation

Figure 6.19 shows the designed trajectory of the target well under investigation. The horizontal well section is in the shale formation and has a 75° inclination angle. The vertical well section

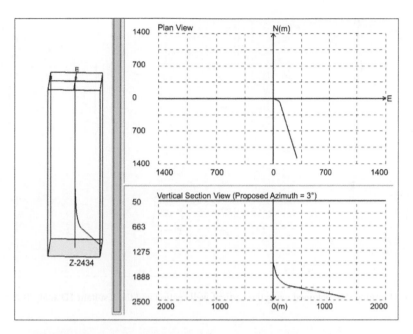

Figure 6.19   The designed trajectory of the target well with a horizontal well section.

was drilled before the horizontal well. Figure 6.20 shows the logging data obtained from the vertical well section that includes gamma ray data (blue curve in first track on the left), sonic data (blue curve in second track on the left), and density logging data (third track on the left). The one-dimensional (1D) geomechanical analysis of the initial geostress solution was performed using these logging data. The following sub-sections describe the analysis results.

6.7.2   1D geomechanics analysis

The first modeling step is to perform a 1D geomechanics analysis. The *PP* obtained using logging data is indicated by the red line in the second track from the right in Figure 6.20. This resultant *PP* is approximately the same as the value of hydrostatic pressure.

6.7.2.1   *Analysis of geostress*
The stress pattern of initial geostress is analyzed, based on natural fractures. The purpose of the stress pattern analysis is to determine the order of the three principal stress components.

6.7.2.2   *Natural fracture and stress pattern*
Figure 6.21 provides information about the natural fractures. Two types of natural fractures are measured: natural fractures with high hydraulic conductivity (*HC*) and natural fractures with low *HC*. Two characteristics are associated with natural fractures with high *HC* in this area. The azimuth direction is northeast (NE) to southwest (SW). The inclination angle of the natural fractures is approximately 80 to 90°. Lower *HC* fractures have similar azimuth directions to those of high *HC*, but their inclination angles are between 40 and 70°.

Figure 6.22 shows Poisson's ratio data. The purple curve is the value of Poisson's ratio in the vertical direction, and the blue curve is Poisson's ratio in the horizontal direction. The vertical value of Poisson's ratio is obviously larger than the value in the horizontal direction for the *TVD* interval above 2000 m. At the *TVD* interval below 2000 m, Poisson's ratio in the vertical direction is near that in the horizontal direction. This indicates the *SH* is near the vertical stress, but still less than the vertical stress.

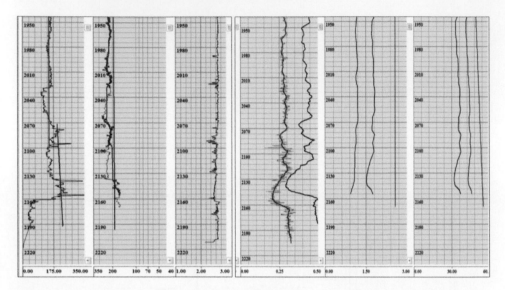

Figure 6.20   Logging data obtained using the vertical well section and the resultant 1D analysis curves.

With reference to the inclination angle of high-*HC* natural fractures, it can be concluded that the *Sh* is the minimum principal stress, and the stress pattern of the formation is regular, with *Sv* being the maximum stress component (i.e., $Sv \geq SH \geq Sh$). Here, the sign convention indicates that the compressive stress is positive.

Natural fracture width values in a shale formation are not shown in image logging data. However, based on Lietard *et al.* (1996) and Araujo *et al.* (2005), 1 mm is used as the width of natural fractures in this shale formation.

6.7.2.3   Calculation of PP and FG

The values of static Poisson's ratio for the *TVD* interval from 1950 to 2190 m are indicated by the red curve (Fig. 6.20, third track from the right). The values of effective stress ratio *k*0 calculated using the values of Poisson's ratio for this *TVD* interval are indicated by the black curve in this track. The maximum value of *k*0 is approximately 0.5, but the minimum value of *k*0 is approximately 0.3, which is significantly low. While drilling the vertical well section, lost circulation was recorded. This indicated the low value of the effective stress ratio *k*0.

PP values obtained from sonic data are indicated by the red curve in the central track. Because of the limitation of sonic data from shale points, the *PP* results are only obtained for the interval above *TVD* = 2160 m.

The *Sh* obtained using the Mathey and Kelly method is indicated by the blue line in the central track of Figure 6.20. *Sh* is also used as the *FG*. The curve of effective stress ratio was used in the calculation of *Sh*.

The minimum value of the window between the *PP* and *FG* is approximately 0.293 kPa/m (2.5 ppg).

The goal of hydraulic plugging is to widen the MWW to 4.067 kPa/m (4.16 ppg) at the interval around *TVD* = 2150 m. This objective is met by increasing the *FG* with the amount of 1.623 kPa/m (1.66 ppg) from its current value. To convert this value of stress gradient to the stress magnitude at *TVD* = 2150 m, an increase in *Sh* to 4.285 MPa is necessary. This accomplishes the task of hydraulic plugging.

The initial *PP* value at this *TVD* point is 27.3 MPa, the value of *Sh* is 32.0 MPa, the *SH* is 42.5 MPa, and the *Sv* is 54.5 MPa.

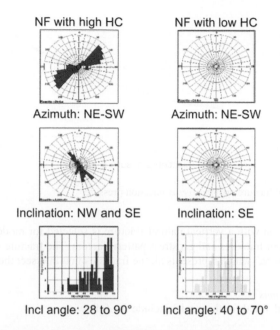

NF with high HC   NF with low HC

Azimuth: NE-SW   Azimuth: NE-SW

Inclination: NW and SE   Inclination: SE

Incl angle: 28 to 90°   Incl angle: 40 to 70°

Figure 6.21   Natural fractures information obtained from image logging data.

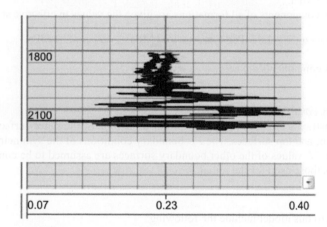

Figure 6.22   Values of Poisson's ratio for the section around the target formation.

Effective stress is used in this calculation. The effective stress component values of the *TVD* at 2160 m are as follows:

- $Sx = Sh = -4.7\,\text{MPa}$
- $Sy = SH = -15.2\,\text{MPa}$
- $Sz = Sv = -27.2\,\text{MPa}$

6.7.3   Hydraulic plugging numerical analysis

The trajectory of the horizontal/high-angle inclined well section is designed to be in the direction of *SH*.

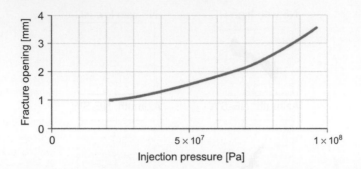

Figure 6.23   Variation of fracture opening *versus* injection pressure.

A vertical well section with a vertical natural fracture is selected for modeling based on the analysis of major natural fractures and geostress pattern factors. The fracture used in this model is the "equivalent fracture," which represents all the fractures that intersect the wellbore surface.

6.7.3.1   *Initial conditions*
The initial conditions defined for the model include the following:

- Void ratio is 0.125.
- Initial PP is 27.3 MPa.
- Initial geostress components of all formations are given the values in effective stress space as:

$$\sigma_x = -4.7\,\text{MPa}; \quad \sigma_y = -15.2\,\text{MPa}; \quad \sigma_z = -27.2\,\text{MPa}$$

All shear-stress components are zero.

6.7.3.2   *Boundary condition*
Displacement constraints are applied to the normal directions of the outer surface, symmetry surface, and bottom and top surfaces. The PP boundary conditions of the symmetry surface are given as zero. The PP values of the other boundary surfaces are assumed to be constant and are the same values as those used for the initial PP.

6.7.3.3   *Loads*
The loads applied to this model include the following:

- Geostress and gravity load.
- Concentrated flow during the injection step.
- Mud-weight pressure applied to the inner surface of the wellbore.

The location of the injection flow application is the same as that presented in Section 6.6.

6.7.3.4   *Material parameter values*
Tables 6.1 and 6.2 list the material parameter values used in the calculation.

6.7.3.5   *Numerical analysis of hydraulic plugging and fracture opening*
Figures 6.23 and 6.24 show the numerical results of fracture opening with injection pressure, which is the mud pressure during hydraulic plugging. The fracture range varies from 1 to 3.5 mm with a variation of mud pressure from 20 to 90 MPa.

Figure 6.24 shows the variation of fracture opening with time.

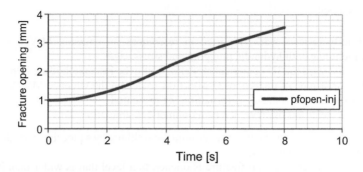

Figure 6.24   Variation of fracture opening with loading history.

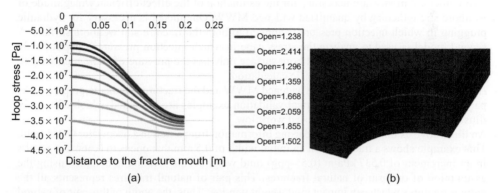

Figure 6.25   (a) Hoop stress distribution around the wellbore at various time moment t along Path 1; (b) location of points of observation in Path 1 (red dots on the line).

6.7.3.6   *Numerical analysis of stress caging effect*

Figure 6.25a shows the hoop stress variations around the wellbore along Path 1 in Figure 6.25b under a defined mud-pressure value. This set of numerical hoop stress distribution results is obtained with the given fracture-opening values of open = 1.238 to 2.414 mm. These fracture-opening values correspond to the results from hydraulic plugging, in practice.

Figure 6.25a shows the hoop stress distribution along Path 1 in Figure 6.25b. The minimum value of magnitude of hoop stress is that of *Sh*. It occurs at the mouth of the fracture.

With a fracture-opening value of open = 1.28 mm, the value of magnitude of effective *Sh* increases from the original 4.7 MPa to approximately 9 MPa. This corresponds to a total *Sh* value of 36.3 MPa. In this case, the effectiveness of caging is 4.3 MPa.

The value of *Sh* for open = 2.424 mm is 35 MPa, which corresponds to a total *Sh* value of 62.3 MPa. In this case, the effectiveness of caging is 30.1 MPa.

Based on the numerical results shown in Figures 6.24 and 6.25, the following were determined:

- The task of widening the MWW with the value of 1.623 kPa/m (1.66 ppg) can be realized by opening the fracture to 1.285 mm with hydraulic plugging (Fig. 6.25a).
- The mud-weight pressure necessary to create this value of a 1.285-mm fracture opening is 39 MPa, which corresponds to a 14.67 kPa/m (15-ppg) mud-weight gradient which corresponds to 1.81-g/cm^3 mud density (Fig. 6.25a).
- The fracture-opening value is determined to be 1.285 mm; therefore, the LCM size is chosen as a mixture from 100 to 500 μm.

6.8   CONCLUSIONS

This chapter briefly reviews works reported in references on widened MWW by wellbore strengthening, including both hydraulic plugging and stress cage.

The workflow and steps for 3D numerical modeling using the FEA method are introduced. Two examples of the application of the proposed workflow to design the widened MWW for drilling in a subsalt well section and a shale gas formation are presented.

The following conclusions are obtained using numerical simulation:

- Hydraulic plugging seals natural fractures around the wellbore by squeezing particles of LCM, thus stopping loss of mud weight.
- Stress cage forms when a natural fracture is opened to a level that is wider than its original width value. When the fracture width increment is significant, the stress cage is strong and can effectively widen the original MWW.
- Two numerical models are necessary for the estimation of the effect/efficiency/magnitude of wellbore strengthening by quantified widened MWW. One model is necessary for hydraulic plugging in which injection pressure is applied to both the fracture and wellbore surface. A second model is used to analyze the stress cage in which injection pressure is applied only on the fracture surface, and the wellbore surface only has the pressure of static mud weight, which is a constant value for a specific calculation.
- A numerical scheme to calculate the widened MWW and the fracture opening to prepare LCM particle sizes is established. Numerical validation examples are presented, which effectively illustrate the efficiency of the proposed workflow.
- An illustrative example is performed using information from a well in a deepwater GOM field. This example shows a fracture with an initial width of 0.5 mm that opens to 0.664 mm results in an increment of 0.537 kPa/m (0.55-ppg) mud weight. This solution is obtained using the assumption of one pair of natural fractures. This pair of natural fractures represents all the natural fractures which can impact mud weight window. Thus, the width of this pair of natural fractures should be derived from image logging data of borehole.

Natural fracture permeability is a key factor that influences the numerical solution for analysis of hydraulic plugging and stress caging. It is a topic for further work.

CHAPTER 7

Numerical estimation of upper bound of injection pressure window with casing integrity under hydraulic fracturing

This chapter briefly introduces the phenomena of casing deformation under hydraulic fracturing injection pressure. The concept and workflow for modeling casing deformation under hydraulic injection pressure is introduced. The upper bound of safe injection pressure window is then obtained by limiting the casing deformation to a specific threshold of injection pressure.

7.1 INTRODUCTION

In the petroleum industry, a well that is not producing as expected may require stimulation to increase the production of subsurface hydrocarbon deposits, such as oil and natural gas. Hydraulic fracturing has long been used as a major technique for well stimulation. The rapid development of unconventional resources in recent years has led to a renewed interest in hydraulic fracturing, and multistage hydraulic fracturing in particular. Examples of such unconventional resources include, but are not limited to, oil and/or natural gas trapped within tight sand, shale, or other type of impermeable rock formation. A multistage hydraulic fracturing operation may involve drilling a horizontal wellbore and applying a series of stimulation injections along the wellbore over multiple stages.

A key factor to the success of such a hydraulic fracturing operation is maintaining ca sing integrity along the wellbore during each stage of the operation. Significant casing deformation in a section of the wellbore can hinder or even stop the progress of the hydraulic fracturing operation altogether. For example, as reported by Shen (2014), such casing deformation may prevent the removal of bridge plugs or other operational work that may need to be performed for that section before the operation can proceed to other sections of the wellbore. Consequently, several well sections or even the entire well may have to be abandoned as a result of casing deformation that could occur before all stages of the hydraulic fracturing operation have been completed.

Therefore, an effective design for a multistage hydraulic fracturing operation should account for the potential casing deformation that could occur during different stages of the operation. Such an effective hydraulic fracturing design can then be used to mitigate the risks of an expensive failure in the casing during the actual operation.

The focus of this work is to establish a workflow with a simplified 3D model for the numerical solution of the maximum value of safe injection pressure of hydraulic fracturing (HF). This resultant injection pressure value can prevent significant deformation to the casing under HF injection load.

In this work, the following contents are described:

- The simplified 3D numerical model that reproduces the measured phenomenon of casing deformation under HF loading of a horizontal well section. Major factors affecting casing deformation can be demonstrated by this numerical model.
- The workflow of using the aforementioned simplified 3D model to predict the maximum value of safe injection pressure which is the upper bound of safe injection pressure window.

This work focuses on casing deformation after the HF operation of horizontal well sections. With reference to the existing documents and observations, the mechanism of casing deformation

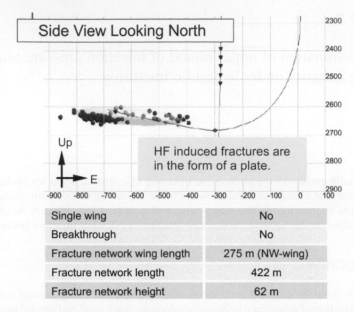

Figure 7.1   Distribution of HD-induced fractures in the formation: asymmetric to the trajectory: side view.

resulting from HF at vertical well sections can be different from the casing deformation at horizontal well sections, and it is not a topic of discussion here. In the following sections, discussions are limited to the topics related to horizontal wells only.

Figures 7.1 and 7.2 show a case (Case 1) of measured microseismic data induced by hydraulic fracturing, along with the well trajectory and locations of casing deformations. These are two typical cases of casing deformation observed by practical engineering in a shale gas field in China. In Case 1 shown in Figure 7.1, significant casing deformation occurred after a HF formation treatment in a horizontal well. Consequently, the work of removing the bridge plug could not be completed for this well. The measured location of the casing deformation is at the range of perforation section of that HF stage. With the measured microseismic data of the HF work shown in Figure 7.1 and Figure 7.2, it is observed that the distribution of HF induced fractures is significantly asymmetrical to the well trajectory. It is thus further derived that the distribution of natural fractures is asymmetrical to the well trajectory.

Figure 7.3 shows the well trajectory of Case 2. The measured casing deformation location is different from that shown in the examples provided in Figure 7.1 and Figure 7.2. In Case 2, the HF-induced casing deformation occurred at the heel of the trajectory.

Figure 7.4 shows the design of the HF stages. Figure 7.4 shows that Stage 11 of the fracturing design is at the location of the heel. The distribution of formation properties such as stillness etc. to the trajectory at the heel is asymmetrical in this section; the formation n thickness above the trajectory increases with measured depth, but is always smaller than thickness in the formation below.

The simplified 3D model designed in this chapter is used for the numerical solution of the maximum safe injection pressure value for hydraulic fracturing. The following sections present a description of the major factors considered in the model. These factors are selected from a group of factors that affect the casing deformation under HF injection.

The workflow for the numerical solution of the maximum safe injection pressure value for hydraulic fracturing is presented after the model description.

A validation example is presented at the end of this section to illustrate the workflow and the model.

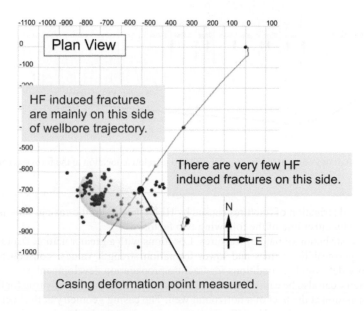

Figure 7.2   Distribution of HD-induced fractures in the formation: asymmetric to the trajectory: plane view.

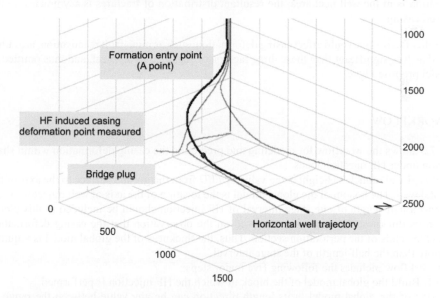

Figure 7.3   Trajectory of the well for Case 2 and the location of casing deformation.

Based on the observation shown in Figure 7.1 through Figure 7.4 and other related references, such as Dassault Systems documentation (Dassault Systems, 2008), the three major factors affecting the casing deformation under HF injection loads are given in the following:

- Injection pressure and/or injection rate. Obviously, without injection, there would be no significant casing deformation.
- Imperfections of cementing rings. Imperfection of cementing results in non-uniform loading to the casing that lies within it. Consequently, it will exacerbate the casing deformation caused by HF injection.

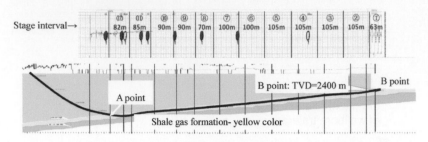

Figure 7.4   Design of trajectory of this horizontal well and its relative location to the formation of shale gas.

- Asymmetrical distribution of fractures caused by HF to the axis of the casing. This asymmetry can be caused by either one of the following two factors:
- Asymmetric distribution of natural fractures. Locations with a greater natural fracture density have higher permeability values and lower formation strength values; consequently, these locations provide favorable conditions for fracture propagation development.
- This asymmetry can also be caused by structural factors. This is particularly true for the case of casing deformation at the heel of a horizontal well. The casing geometry at the heel section is curved. If the fractures induced by HF injection are very near the heel, or even if the fracturing operation is in the well heel area, the resultant distribution of fractures is asymmetric to the curved casing.

The other factors that could affect casing deformation, such as mini-fault reactivation, are either regarded as less significant than these three factors, or regarded as atypical, and thus omitted in the model proposed here.

7.2   WORKFLOW

Figure 7.5 shows the workflow for modeling and predicting the casing deformation within shale formation under HF injection loads.

For this global model, its height is the value of *TVD* from the ground surface to the axis of the horizontal well. The bottom of the global model is the location of the axis of well trajectory sites.

At least a half-length of one stage interval of HF injection should be included in this global model in its thickness direction. This is based on the observation that the casing deformation begins at the ends of the perforation section. Thus, the thickness of the global model is required to be more than the half-length of the stage interval.

This workflow includes the following five major steps:

Step 1: Build the global model of the block in which the HF injection is performed.

The size of the global model in its length direction can be any value between the range of 30 times the diameter of the wellbore and twice the half length of the gene rated fracture of HF injection. The size used in the validation example is 120 m. With reference to the Saint-Venant principle, stress variation away from the casing axis in the lateral direction has little effect on the deformation of casing. Therefore, it is necessary to have a large model, but it is not necessary to make it as large as the induced fracture length. The lower part of the global model is for the formation that is to be hydraulically fractured. The upper part of the global model is for the overburden formations. Details about formation geomechanical properties (such as inclination of bedding planes, thickness of each formation layer, and faults) and density, Young's modulus, and Poisson's ratio should be modeled as much as possible. However, because no porosity exists in these overburden layers, nonpermeable material models should be used for the purpose of model simplification.

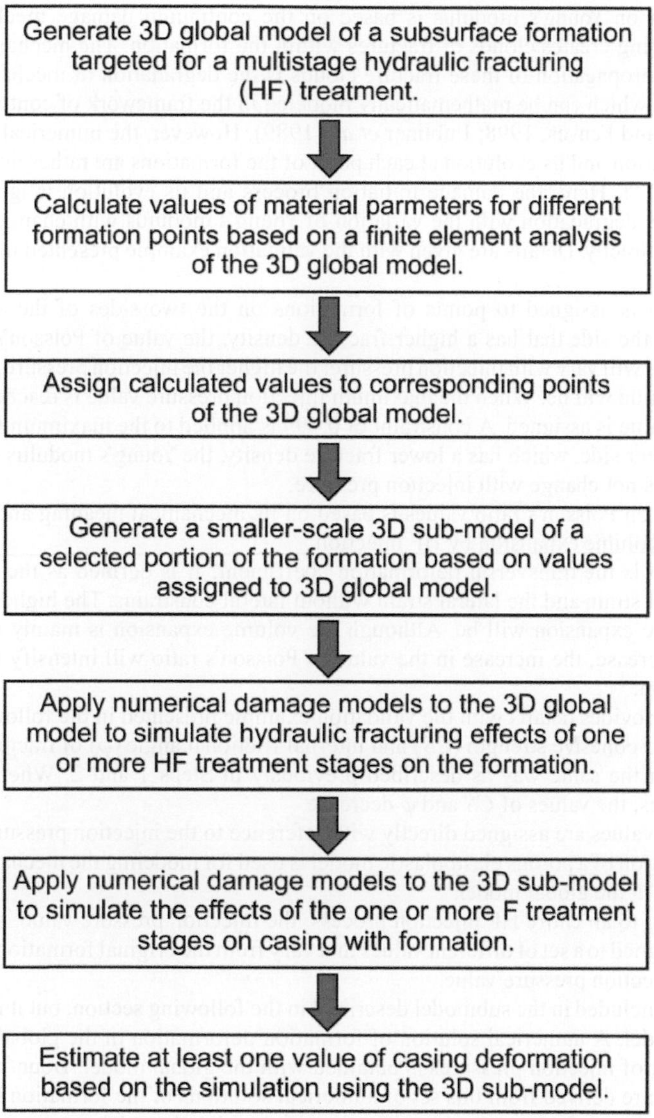

Generate 3D global model of a subsurface formation
targeted for a multistage hydraulic fracturing
(HF) treatment.

⬇

Calculate values of material parmeters for different
formation points based on a finite element analysis
of the 3D global model.

⬇

Assign calculated values to corresponding points
of the 3D global model.

⬇

Generate a smaller-scale 3D sub-model of a
selected portion of the formation based on values
assigned to 3D global model.

⬇

Apply numerical damage models to the 3D global
model to simulate hydraulic fracturing effects of one
or more HF treatment stages on the formation.

⬇

Apply numerical damage models to the 3D sub-model
to simulate the effects of the one or more F treatment
stages on casing with formation.

⬇

Estimate at least one value of casing deformation
based on the simulation using the 3D sub-model.

Figure 7.5   Flowchart of the workflow for casing deformation calculation under HF injection.

Step 2: Assign values to material parameters of mechanical properties of the global model, particularly to the portion of hydraulically fractured formation.

To simulate the asymmetrical distribution of the fractures generated by the HF, different values of parameters of mechanical properties are assigned. The following paragraphs provide information about this data assignment.

Young's modulus is assigned to points of formations on the two sides of the axis of casing respectively. For the side with a higher fracture density, the Young's modulus value at each point of the model varies with the injection pressure: the higher the injection pressure, the lower the value of Young's modulus will be. When the maximum injection pressure value is reached, the lowest value of Young's modulus is assigned. For the other side, which has a lower fracture density, the Young's modulus value remains constant and does not change with injection pressure.

This operation on Young's modulus is based on the continuum damage mechanics theory. Hydraulic fracturing creates clouds of fractures within the formation. The mechanical effect on the creation and propagation of these fracture clouds is the degradation of mechanical stiffness of the formation, which can be mathematically modeled in the framework of continuum damage mechanics (Lee and Fenves, 1998; Lubliner *et al.*, 1989). However, the numerical simulation of the damage initiation and its evolution at each point of the formations are rather time-consuming (Shen, 2012, 2014). Here, the damage initiation process and its evolution is ignored, and the resultant stiffness degradation with the variation of Young's modulus with changes of injection pressure is used directly. Details are given with the validation example presented in the following sections.

Poisson's ratio is assigned to points of formations on the two sides of the axis of casing respectively. For the side that has a higher fracture density, the value of Poisson's ratio at each point of the model will vary with injection pressure: the higher the injection pressure, the higher the Poisson's ratio value will be. When the maximum injection pressure value is reached, the highest Poisson's ratio value is assigned. A constraint of 0.499 is applied to the maximum Poisson's ratio value. For the other side, which has a lower fracture density, the Young's modulus value remains constant and does not change with injection pressure.

The operation on Poisson's ratio values is based on its mechanical meaning and the observed phenomenon of volume expansion by HF injection.

Poisson's ratio is the transversal deformation coefficient. It is defined as the negative ratio between the axial strain and the lateral strain without lateral constrains. The higher its value, the larger the volume expansion will be. Although the volume expansion is mainly a result of the pore pressure increase, the increase in the value of Poisson's ratio will intensify the amount of volume expansion.

This chapter provides details with the validation example presented in the following sections.

Degradation of cohesive strength (*CS*) and internal frictional angle (φ) of fractured formation are performed in the same way as described previously in Steps 1 and 2. When the injection pressure increases, the values of *CS* and φ decrease.

Pore pressure values are assigned directly with reference to the injection pressure variation. A coupled formulation of a porous elastoplastic model is used for modeling the mechanical behavior of the formation in the global model.

Corresponding to an entire HF injection process, the injection pressure value in the fractured formation is assigned to a set of different values that vary from the original formation pore pressure to the highest injection pressure value.

The casing is included in the submodel described in the following section, but it is not included in the global model. A numerical solution of formation deformation in the global model corresponding to a set of injection pressures is obtained with the global model. Boundary conditions of the submodel are derived from this set of numerical solutions of the formation's deformation.

Model calibration and parameter identification of the global model can be performed with measured data. These data sets can be either of the following: the value of casing deformation measured downhole or the value of measured ground surface deformation.

Step 3: Build the submodel model comprising the cementing ring and casing along with formations under HF injection load.

The geometry of the submodel has the following characteristics:

- The size of the submodel is 20 m in length and 10 m in width and height, respectively.
- The casing and cementing ring are modeled in the submodel.
- The bottom of the submodel is on the bottom of the global model.
- The center of the bottom of the submodel is at the same location as that of the global model.

The symmetry property is assumed of the casing: only the upper part of the casing and its surrounding cementing are modeled in this submodel. Deformation behavior of its lower part is regarded as the same as this upper part.

Material properties comprised in the submodel have the following characteristics:

- The material properties of formations included in the submodel are the same as those of its corresponding parts in the global model.
- Cementing material is included in the submodel only. To model the factor of poor cementing quality, the cementing ring in this submodel includes two kinds of cementing material:
  - Cementing material CM-1 represents good quality of cementing, and thus its parameters of mechanical properties have normal values as given in various manuals of cementing operation.
  - Cementing material CM-2 represents poor quality of cementing, and thus its parameters of mechanical properties have smaller values compared to those used for CM-1.

A special case is to assign the same value to material parameters for both CM-1 and CM-2 to simulate the mechanical behavior of the model with good cementing quality.

The following sections provide details with the validation example presented.

The casing material is steel, and the elastoplastic model with a linear strain hardening property is adopted.

Boundary conditions of the submodel are taken from the numerical solution of the global model described previously.

Step 4: Perform calculations with the global model.

Step 5: Perform calculations with the submodel.

With the numerical solution of casing deformation under the given set of values of injection pressure, the safe maximum injection pressure value can be determined: it is the value of injection pressure before the occurrence of significant casing deformation.

Taking this safe value of maximum injection pressure as reference in the design of a HF injection operation, the highest efficiency of HF operation can be achieved with confirmed casing integrity.

7.3   VALIDATION EXAMPLE

This validation example is taken from an actual case of casing deformation in an HF injection in Sichuan Basin, southwest China (Shen, 2014). The data values used in the modeling are a variation of the original project data. This example is for workflow illustration purposes only.

The original design of the HF operation includes 10 stages of HF stimulation. Casing deformation occurred after the HF operation in the third stage. Figure 7.1 shows the measured micro-seismic data. Figure 7.6 shows the injection pressure recorded during the injection operation of this stage.

The input data of the model include the following:

- Initial geostress field: sequence and direction of principal stress.
- Casing: geometric parameters, material parameters.
- Cement sheath: geometric parameters, material parameters.
- Mechanical properties of the rock formations.
- Injection pressure.

7.3.1   **Initial pore pressure**

The specific values of the parameters are provided in the following sections.

7.3.2   **Initial geostress field: sequence and direction of principal stress, and initial pore pressure**

The *TVD* of the casing is 2600 m, the vertical stress is $Sv = 63$ MPa, the minimum horizontal principal stress is $Sh = 66.2$ MPa, and the maximum horizontal principal stress is $SH = 66.6$ MPa.

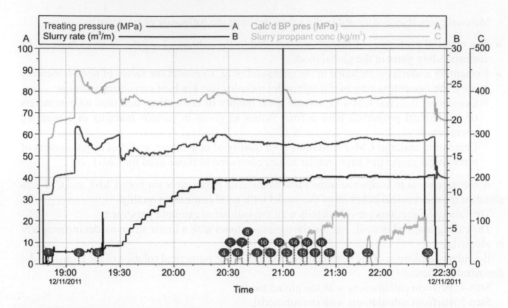

Figure 7.6   The injection pressure during the injection operation of this stage: green line, bottomhole.

The direction of the maximum horizontal principal stress (*SH*) is parallel to the wellbore axis. The initial pore pressure is set as 30 MPa.

A reverse fault stress pattern was found in practice for geostress distribution in this region.

7.3.3   Casing: geometric parameters, material parameters

The inner diameter of the casing is 0.1214 m, the casing thickness is 0.0091494 m, the material density of the P110 steel of which the casing is made is 7922 kg/m^3, and the initial yielding is 758 MPa. The modulus of elasticity is $E = 206$ GPa, the modulus of shearing is $G = 79.38$ GPa, and the Poisson's ratio is 0.3. In the calculation, the ideal elastoplastic model is used to simulate plastic deformation (Dassault Systems, 2008) of the casing material.

7.3.4   Cement ring: geometric parameters, material parameters

The inner diameter of the cement sheath is 0.1397 m, the outer diameter is 0.2159 m, the material density is 1900 kg/m^3, the modulus of elasticity of regular cementing material is $E = 27.2$ GPa, and the Poisson's ratio is 0.3. To simulate the unevenness of the cement sheath filling caused by poor well cementation quality, the dark (Cem-1) and light (Cem-2) parts of the cement sheath material in the figure are given different modulus values of elasticity respectively. In detail, Young's modulus given to Cem-1 is the original typical value of $E = 27.2$ GPa. The value for Cem-2, which is the weaker one, is $E1 = 10\%$ of $E = 2.72$ GPa.

7.3.5   Mechanical properties of the rock formations

The rock density used in the model is 2650 kg/m^3, the modulus of elasticity is $E = 40$ GPa, and the initial value of Poisson's ratio is 0.25. To simulate the unevenness of the natural fracture distribution, the green and red parts of the formation material in Figure 7.7 are given different modulus values of elasticity, respectively.

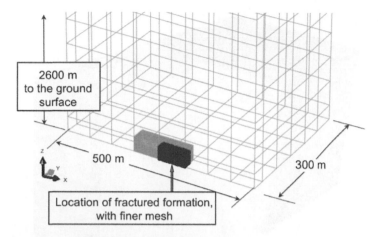

Figure 7.7   Illustration of the FEM mesh of the global model.

7.3.6   Stiffness degradation

In the work reported, the maximum value of Young's modulus E degradation within the formation around the wellbore is 30%. Accordingly, values of Young's modulus used in this model are $E0$ for formation-1 and 70% of $E0$ for E of formation-2.

7.3.7   Injection pressure

The bottomhole injection pressure of the fracturing process is calculated from pumping pressure with the assumption of no friction drag. The peak value of fluid pressure applied on the inner surface of the casing is $P = 90$ MPa.

Injection pressure values are assigned to the fractured formation as its pore pressure in the process of HF injection.

The formation that is not successfully fractured will retain its original value of pore pressure.

7.3.8   Boundary conditions to the global model

The boundary conditions of the model are set up as follows: zero displacement constraint in the normal direction on both lateral surfaces, as well as the bottom surface. The top surface is free of load and displacement constraints, which simulates the ground surface.

7.3.9   Finite element mesh of the global model

Figure 7.7 shows the location of the fractured formation. To have a higher displacement solution accuracy, the mesh density for the part of the model in the neighborhood of the fractured formation is much denser than that of the other parts.

The trajectory of the casing is in the Y-direction, and the X-direction is the direction in which the HF induced fracture propagates. The X-direction is thus the direction of maximum horizontal stress.

7.3.10   Finite element mesh of the submodel

The submodel shown in Figure 7.8 illustrates the location of the fractured formation. It takes 1/4 of the model geometry. The casing and cement ring are included in the submodel.

Figure 7.9 shows the details of modeling the imperfections of the cementing ring with the submodel in a cross-sectional view.

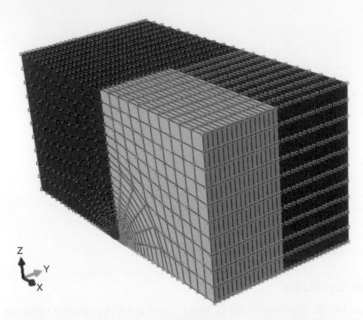

Figure 7.8   Illustration of the FEM mesh of the submodel.

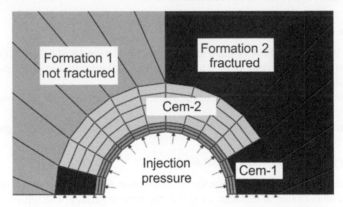

Figure 7.9   Illustration of modeling the imperfections of the cementing ring with the submodel: cross-section view.

The cementing ring is divided into three parts in Figure 7.9.

The part Cem-2 in the center of the cross-section of the cementing ring in Figure 7.8 is assigned as the weak portion of the cementing ring, which represents the imperfection of cementing.

The other two parts of Cem-1 are assigned as normal cementing parts, which represent good quality cementing work.

For the case of good quality cementing work, the values of parameters, such as Young's modulus, are assigned their normal values as designed.

For the case of poor quality cementing work, the value of Young's modulus of weak cementing can be assigned a significantly smaller value, even down to zero value. In the case of the example, 10% of the typical cementing Young's modulus value is used as the value for this weak section.

Cementing quality that is between good and poor can be modeled with parameter values between the previously described values. This is easy to achieve by assigning a value directly in the numerical model.

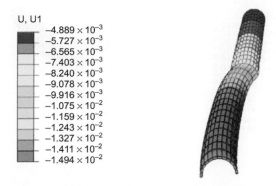

Figure 7.10   Casing deformation with good cementing quality: deformed mesh.

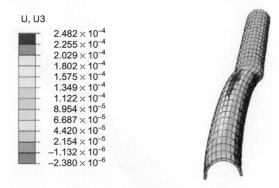

Figure 7.11   Casing deformation with poor cementing quality: deformed mesh.

In addition, the percentage of area of the weak part of the cementing ring can be varied with reference to the measured quality index of the cementing work.

7.3.11   Numerical results of casing deformation

Details of the calculation of the global model are not addressed here.

Figure 7.10 and Figure 7.11 show the deformed mesh of the casing under HF injection pressure with good quality of cementing. $U1$ is the value of deformation in the horizontal lateral x-direction, and $U3$ is the value of deformation in the vertical direction.

Figures 7.12 and 7.13 show the deformed mesh of the casing under HF injection pressure with poor quality of cementing.

Comparisons of the casing deformation in Figure 7.10 through Figure 7.13 show that the maximum values of lateral deformation $U1$ are similar for these two cases. However, the maximum vertical displacement values of the casing deformation for the case of poor cementing quality are significantly larger than for those good cementing quality. Consequently, it is concluded that poor cementing quality intensifies casing deformation.

Table 7.1 presents a set of numerical results of casing deformation along with the specified injection pressure value at the moment. The casing deformation value is the maximum value of displacement component that occurs in the casing.

From Table 7.1, the following conclusions can be drawn:

- For the case of 90 MPa peak value of injection pressure, the cement quality significantly affects the vertical displacement value, but has little influence on the lateral displacement

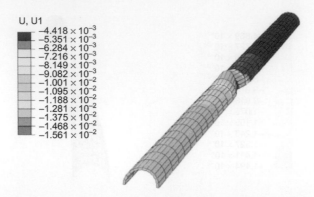

Figure 7.12   Casing deformation with poor cementing quality: deformed mesh.

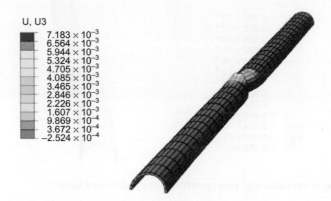

Figure 7.13   Casing deformation with poor cementing quality: deformed mesh.

Table 7.1   Maximum values of von Mises equivalent stress within the casing corresponding to each set of input data.

Injection pressure [MPa]	Maximum value of casing deformation [mm]		Distribution of asymmetric properties	
	Lateral	Vertical	Fractures	Cement ring quality
90	15.6	7.18	asymmetric	poor
90	14.9	0.2	asymmetric	good
80	9.5	0.83	asymmetric	poor
80	5.7	0.34	asymmetric	good

value. Consequently, ovalization of the casing's cross-sectional surface is mainly caused by poor cementing quality.
- For the case of less injection pressure of 80 MPa, displacement values in both lateral and vertical directions are significantly less than their values with the 90 MPa injection pressure. In this case, the poor quality of the cementing ring has less effect than 90 MPa injection pressure.
- 80 MPa is a safe maximum injection pressure value. This value as the upper bound of injection operation helps to ensure the casing integrity.

Therefore, it can be concluded that the significant casing deformation that hinders successful stimulation is caused by the joint action of the three factors listed previously:

- High value of injection pressure.
- Asymmetric distribution fractures, both natural fracture and induced fractures.
- Poor quality cementing ring.

7.4   ENDING REMARKS

Efficiency and accuracy of the proposed model and workflow are the major commercial competitive advantages of this chapter. The simplified 3D numerical model established in this chapter models only important essential mechanical characteristics included in the phenomenon of casing deformation under HF injection, and omits other details that could result in a significant increase of computational burden with little effect on the casing deformation. This helps to ensure the efficiency of the model and workflow proposed here.

The proposed model and workflow capture the major aspects of mechanical behavior of the coupled deformation of the formation and the porous flow related to the HF Injection. A series of values of injection pressure and formation pore pressure are assigned to the model in the analysis. Interaction between the formation and the casing are calculated in a fully coupled way. This helps to ensure the accuracy of the numerical solution of casing deformation under HF injection loading obtained with the numerical model and workflow developed here.

The validation example illustrates the accuracy of the numerical solution of the casing deformation; it also proves the efficiency of the numerical analysis workflow.

The efficiency and accuracy of workflow is proven using the previously described simplified 3D numerical model to predict the maximum value of safe injection pressure.

CHAPTER 8

Damage model for reservoir with multisets of natural fractures and its application in the simulation of hydraulic fracturing

The presence of natural fractures can significantly affect the quality of hydraulic fracturing operations in tight-sand and shale oil/gas formations. This chapter describes the concept and workflow used to model natural fractures with continuum damage tensor and the resulting orthotropic permeability tensor. A hydraulic fracturing model example that uses damage variables is presented. In this example, a damage-dependent permeability tensor is proposed in tabular form. The issue caused by the existence of an angle between directions of principal stresses and that of natural fractures-related damage tensor are solved by introducing local directions in the model. For the purpose of comparison, an example of hydraulic fracturing is provided at the end of the chapter for the case that no angle exists between the directions of principal stresses and that of natural fractures. Contents presented in Chapters 1 through 3 are referenced.

8.1 INTRODUCTION

Natural fractures and their role in the modeling of petroleum reservoirs have been investigated by various researchers for more than 50 years. Sufficient reference articles exist. Natural fractures in reservoir formations can significantly affect stimulation- induced fracture propagation (Carnes, 1966; Parvizi *et al.*, 2015; Potluri *et al.*, 2005; Rodgerson, 2000; Shahid *et al.*, 2015). Here, natural fracture concepts and properties are briefly reviewed. The various types of natural fractures include the following:

- Joint: natural fracture opening (Mode I crack).
- Fault: natural fracture caused by shear sliding that is usually closed (Mode II crack and Mode III crack).
- Stylolite: natural fracture caused by compressive deformation that in the normal direction is usually closed.

The geometrical properties of natural fractures usually include the following:

- Aperture: the opening of a fracture.
- Spacing: distance between two neighboring fractures; spacing sometimes alternatively represented by density of fractures.
- Density: number of fractures included in a unit length, sometimes alternatively represented by spacing between two neighboring fractures.
- Length and width: form the area of a fracture.

Geometrical properties of natural fractures can be obtained directly or indirectly. Direct measures to obtain this information include core test and measurement in a laboratory, video imaging, and borehole camera. Indirect measures include various logging measures, seismic data, and inverse analysis by means of well production testing.

Detection devices have limited resolution. Therefore, a direct method (e.g., image camera) can only obtain information from the partial natural fractures that are of sufficient size. The smaller fractures are ignored by these devices. However, natural fracture information obtained using an indirect method (e.g., inverse analysis) is accurate because it matches the phenomena. In general, natural fracture information obtained by inverse analysis is of an equivalent natural fracture system. It does not exactly fit the fracture density and aperture, respectively. It fits the phenomena as a whole.

Natural fracture properties used during modeling of porous flow include the following:

- Porosity.
- Permeability.
- Connectivity.

The porous flow properties can be derived from various logging data (sonic log and neutron density log). These properties can also be derived by inverse analysis using well production testing-flow and pressure tests. Various models are available that measure the porosity and permeability of reservoirs with natural fractures based on the permeability values within the matrix and natural fractures:

- Single porosity and single permeability.
- Dual porosity and single permeability.
- Dual porosity and dual permeability.

The formation matrix of a reservoir with natural fractures (i.e., fractured reservoir) is cut into small blocks by these fractures. The fractures have high permeability, but the porosity of the fractures is relatively low. Matrix permeability is low, but its porosity is relatively high. A dual-porosity and dual-permeability reservoir model provides better accuracy than the single-porosity and single-permeability models.

The dual-porosity and single-permeability reservoir model is a simplified model for dual porosity and dual permeability. This model considers the matrix porosity as a measure for storage only, not for porous flow. Porous flow exists only within the fractures.

A single-porosity and single-permeability reservoir model can further simplify fractured reservoirs (Narr *et al.*, 2006). In this case, porosity and permeability values are determined for the equivalent reservoir model that represents their functions of porosity and permeability in the matrix and fractures as a whole.

This chapter presents the single-porosity and single-permeability model for a reservoir with natural fractures in tight formations. It can model a tight-sand formation with fractures, or shale oil/gas formations with natural fractures. The permeability value is for the reservoir that consists of matrix and fractures as a whole.

The continuum-damage method is an effective measure for the simulation of fracture development caused by stimulation injection. Chapter 1 discusses the concepts and details of continuum-damage mechanics. This work establishes a numerical procedure for the calibration of a damage model using measured information of natural fractures multisets. Image logging data is used. The synthetic value of the damage variable is derived from the work-equivalence principle for the modeling of natural fracture multisets. An example of the procedure illustration is presented.

8.2   EXPRESSION OF NATURAL FRACTURES WITH CONTINUUM-DAMAGE VARIABLE

Geometrical information of a set of natural fractures is usually characterized by their volumetric density, azimuth angle, and inclination angle. Furthermore, fracture properties related to porous flow can be characterized by aperture, permeability, and porosity.

A vector form is used to describe natural fractures. A second-order tensor form is used here to describe the continuum-damage variable.

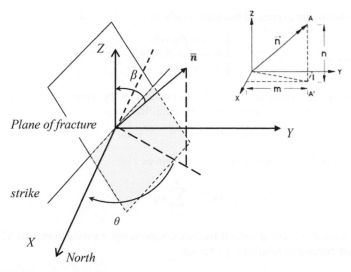

Figure 8.1   Illustration of the spatial relationship between these angles and direction components of the vector of natural fracture.

The second-order tensor is used because the tensor form damage variable is easy to connect with orthotropic permeability. The principal orthotropic permeability value has a unique holonomic relationship with the principal damage value in its related direction.

For one set of natural fractures, the damage tensor comprises one principal value ω and one principal directional vector $\mathbf{n}$. This is the same direction vector as the natural fracture. The principal value of the magnitude of continuum-damage tensor ω is determined by the volumetric density value of the natural fractures ψ. Various versions define natural fracture density (e.g., number per foot or fracture area per unit volume) because of the constraints of the various measuring methods. Therefore, the unit of fracture density ψ is not the same for each version. Consequently, there is no direct explicit relationship between the value of the continuum-damage variable ω and the fracture density ψ. However, there is a functional relationship between ω and ψ, which is defined as f(ψ) in Equation (8.1):

$$\omega = \mathrm{f}(\psi) \tag{8.1}$$

The f(ψ) should be determined by a phenomena-match method that uses either reservoir pressure data measured in the field or in specific porous flow testing data measured in the laboratory.

The direction of a set of natural fractures can be described by a unit directional vector $\mathbf{n}$, which is also used as a principal directional vector of damage tensor. The direction of this vector $\mathbf{n}$ has three directional components (l, m, n). Therefore, the damage tensor in terms of vector of coordinate basis $(\mathbf{i}, \mathbf{j}, \mathbf{k})$ is defined in Equation (8.2) as:

$$\boldsymbol{\omega} = \omega \mathbf{n} \otimes \mathbf{n}, \quad \text{where } \mathbf{n} = l\mathbf{i} + m\mathbf{j} + n\mathbf{k} \tag{8.2}$$

Equation (8.2) can be rewritten in component form as:

$$\omega_{ij} = \omega n_i n_j, \quad i = 1, 3; \ j = 1, 3 \text{ for 3D problem} \tag{8.3}$$

In practice, these directional components (l, m, n) can be calculated in terms of the values of azimuth angle, inclination angle, and x-direction of the coordinate system adopted in a calculation. Figure 8.1 shows the spatial relationship between these angles and directional components of the vector, which represents the natural fractures set. In Figure 8.1, θ is the azimuth angle, β is the inclination angle, and the x-direction of the coordinate system is taken in the north direction.

These relationships are expressed mathematically in Equation (8.4):

$$\begin{cases} l = \sin\beta\cos\theta \\ m = -\sin\beta\sin\theta \\ n = \cos\beta \end{cases} \tag{8.4}$$

The synthetic damage tensor $\boldsymbol{\omega}_T$ equals the summation of the damage tensors for each single natural fracture when there are multiple sets of natural fractures:

$$\boldsymbol{\omega}_T = \boldsymbol{\omega}_1 + \boldsymbol{\omega}_2 + \cdots + \boldsymbol{\omega}_N \tag{8.5}$$

Equation (8.5) can be rewritten in component form as:

$$\omega_{ij}^{(T)} = \sum_{r=1}^{N} \omega_{ij}^{(r)} \tag{8.6}$$

where N is the number of sets of natural fractures, superscript r varies from 1 to N, superscript (T) indicates components of total damage tensor.

8.3   DAMAGE INITIATION CONDITION

The damage initiation condition of the model used here is similar to the Mazars' model described in Chapter 1. For convenience of reading, it is rewritten in Equation (8.7):

$$f_t = \varepsilon_t - Y_t \leq 0 \tag{8.7}$$

The calculation of the damage initiation condition using Equation (8.7) is performed on each of the principal directions of the orthotropic damage model. Subscript t represents tension. Only the case of strain in tension is accounted for in the calculation of damage initiation here.

Chapter 1 defines the variables included in Equation (8.7).

8.4   DAMAGE EVOLUTION LAW

The damage evolution law used here is a holonomic damage model in the form of total quantity other than incremental form:

$$\omega = 1 - Y_{t1}/Y_t \tag{8.8}$$

where Y_{t1} is the maximum value of the damage conjugate force Y, and it corresponds to the maximum fracture opening in principle.

The primary advantage of the Mazars' holonomic damage model is that the damage variable value can be calculated for a given strain status. Iteration at the local level of a material point is not necessary. The primary disadvantage is that numerous material parameters should be calibrated with experimental data.

8.5   DAMAGE-DEPENDENT PERMEABILITY

In this work, a damage-dependent permeability model is used. In general, the value of permeability K_i depends on the value of the continuum-damage variable ω_i:

$$K_i = K_i(\omega_i) \tag{8.9}$$

where subscript i represents the i-th component of orthotropic principal direction, and $i = 1, 2, 3$ for a three-dimensional (3D) problem. There is no summation operation for subscript i in Equation (8.9). K_i represents the value of permeability in the i-th direction of an orthotropic permeability model. ω_i is the variable of continuum damage in the same i-th direction as K_i's. ω_i is also interpreted as the component of damage tensor $\boldsymbol{\omega}$ in its i-th principal direction.

In calculation, the damage-dependent permeability model is introduced through tabular form of a data series, which are calibrated by using the phenomena-match method.

These three orthotropic principal directions of permeability are determined by the directions of the natural fracture, accordingly.

8.6   VALIDATION EXAMPLE: HYDRAULIC FRACTURING OF FORMATION WITH NATURAL FRACTURES

The procedure and models proposed in previous paragraphs are used here to determine the initial value of the continuum-damage tensor of a tight-sand oil reservoir with three sets of natural fractures. Damage tensor and orthotropic permeability are further applied to simulate hydraulic fracturing of a zipper fracture of two horizontal wells in a tight-sand oil formation. Procedures introduced in Chapters 2 and 3 are adopted. Here, the authors only present the model of hydraulic fracturing of the formation and numerical results of fracture generation and propagation represented by the damage variable.

8.6.1   Geometrical information of natural fractures

Three sets of natural fractures were identified from image logging data of the given fractured tight-sand reservoir formation (Fig. 8.2).

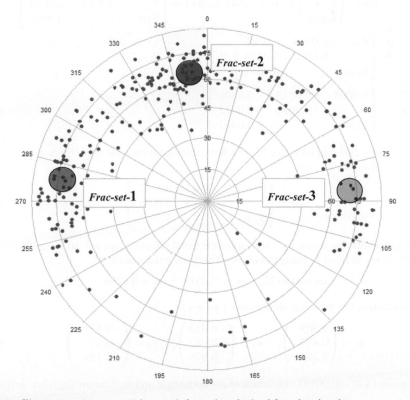

Figure 8.2   Illustration of the natural fracture information obtained from logging data.

Table 8.1   Geometrical information of natural fractures.

	Azimuth angle α [°]	Inclination angle β [°]	Fracture density [1/m]	Aperture [mm]
Frac Set 1	275	72	1	0.5
Frac Set 2	350	68	0.6	0.5
Frac Set 3	80	70	0.4	0.5

Table 8.1 provides the geometrical information of these sets of natural fractures. The values listed in Table 8.1 are for illustrative purposes only. The values are based on data from a tight-sand gas formation and are modified for simplicity purposes.

8.6.2   Damage tensor calculated using natural fracture information

Equation (8.10) defines the principal damage variable values for these three sets of natural fractures as:

$$\omega_1 = 0.015, \quad \omega_2 = 0.03, \quad \omega_3 = 0.01 \tag{8.10}$$

The values of principal damage given in Equation (8.10) are obtained by empirical method. These values should be calibrated with phenomena match by using observed microseismic data related to hydraulic fracturing operations.

With reference to the Equations (8.2) to (8.4) and (8.10), damage tensors for the Set 1 to Set 3 fractures are calculated in Equations (8.11) to (8.13):

$$\boldsymbol{\omega}_1 = 0.015 \begin{pmatrix} 0.083 \\ 0.947 \\ 0.309 \end{pmatrix} \begin{pmatrix} 0.083 & 0.947 & 0.309 \end{pmatrix} = \begin{bmatrix} 0.0001 & 0.0012 & 0.0004 \\ & 0.0135 & 0.0044 \\ \text{Sym} & & 0.0014 \end{bmatrix} \tag{8.11}$$

$$\boldsymbol{\omega}_2 = \begin{bmatrix} 0.025 & 0.0044 & 0.0102 \\ & 0.0008 & 0.0018 \\ \text{Sym} & & 0.0042 \end{bmatrix} \tag{8.12}$$

$$\boldsymbol{\omega}_3 = \begin{bmatrix} 0.0003 & -0.0015 & 0.0006 \\ & 0.0086 & -0.0032 \\ \text{Sym} & & 0.0012 \end{bmatrix} \tag{8.13}$$

In Equations (8.2) through (8.10), the synthetic damage tensor for these three sets of natural fracture has the following value:

$$\boldsymbol{\omega}_{\text{T}} = \begin{bmatrix} 0.0255 & 0.0041 & 0.0112 \\ & 0.0228 & 0.0031 \\ \text{Sym} & & 0.0068 \end{bmatrix} \tag{8.14}$$

A set of direct solution of principal values and principal directions are solved using the synthetic damage tensor $\boldsymbol{\omega}_{\text{T}}$ defined in Equation (8.14). The principal values are:

$$\omega_1 = 0.0332, \quad \omega_2 = 0.0015, \quad \omega_3 = 0.0204 \tag{8.15}$$

Correspondingly, the principal directions of $\boldsymbol{\omega}_{\text{T}}$ are:

$$\mathbf{n}_1 = \begin{pmatrix} 0.808 \\ 0.437 \\ 0.3945 \end{pmatrix}, \quad \mathbf{n}_2 = \begin{pmatrix} -0.415 \\ -0.052 \\ 0.908 \end{pmatrix}, \quad \mathbf{n}_3 = \begin{pmatrix} -0.4716 \\ -0.897 \\ -0.139 \end{pmatrix} \tag{8.16}$$

In Equations (8.4) and (8.16), the values of azimuth angle and inclination angle are defined as:

$$\theta_1 = 61.56°, \quad \beta_1 = 66.76°, \quad \theta_2 = 277.89°, \quad \beta_2 = 24.77°, \quad \theta_3 = 335.06°, \quad \beta_3 = 82.11° \tag{8.17}$$

These three principal directions are orthotropic in 3D space. These values are the initial damage derived from the natural fractures information.

The directions defined in Equation (8.17) will be used as the principal directions of orthotropic permeability tensor.

8.6.3 Numerical simulation of hydraulic fracturing of a formation with natural fractures

8.6.3.1 *Simplification of orthotropic permeability and damage tensor and directions of principal stress*

The simulation of hydraulic fracturing of a formation with natural fractures is not a simple task. Orthotropic permeability caused by multisets of natural fractures introduces orthotropic properties to the entire numerical model. In general, principal stress directions are not the same as the initial permeability tensor derived from natural fractures.

In this case, a global system of numerical model coordinates should be selected from the two, along either the principal stress direction or the principal orthotropic permeability tensor direction.

For the brevity, the values of the initial damage tensor parameters used in Equation (8.18) are modified to the following values:

$$\omega_1 = 0.0132, \quad \theta_1 = 73°, \quad \beta_1 = \beta_2 = 90°; \quad \omega_2 = 0.0015, \quad \theta_2 = 163° \quad (8.18)$$

Thereby, the chosen numerical model coordinates can be in the same principal direction as the initial damage tensor and initial orthotropic permeability tensor.

Next, the direction of minimum horizontal stress (Sh) is assumed at an azimuth angle $\alpha_1 = 90°$. Therefore, there is a 73° angle between the directions of principal stresses and the orthotropic permeability tensor.

The damage initiation value is set as:

$$Y_t = 0.0005 \quad (8.19)$$

The maximum value of damage conjugate force Y_{t1} for the case of damage reaches 1 is given as:

$$Y_{t1} = 0.015 \quad (8.20)$$

Figure 8.3 illustrates the damage dependency of permeability. This diagram is calibrated using the observed phenomena of microseismic data during hydraulic fracturing of offset wells. Both the vertical and horizontal axes in Figure 8.3 are shown in logarithm form.

8.6.3.2 *Finite element model (FEM): mesh, boundary condition, and initial conditions*

The FEM shown in Figure 8.4 is 200 m in both width and length and 25 m in height. A 20-m thick reservoir is covered with a 5-m nonpermeable overburden layer. Injection loading is a point flow in the center of the model. Figure 8.4 shows a 73° angle between the principal directions of permeability tensor and x-axis. The direction of the x-axis is taken in the direction of minimum horizontal principal stress. The geostress level is set for a value as self-weight at true vertical depth $(TVD) = 1600$ m, and it is described with effective stress. Equations (8.21) and (8.22) define the initial conditions.

The following values of stress components are defined for cap rock:

$$S_v = -36.368 \, \text{MPa}, \quad S_h = -28.59 \, \text{MPa}, \quad S_H = -31.66 \, \text{MPa} \quad (8.21)$$

Values of effective stress components, pore pressure, and void ratio in the reservoir are:

$$S_v = -15.7 \, \text{MPa}, \quad S_h = -7.93 \, \text{MPa}, \quad S_H = -11 \, \text{MPa}, \quad P_p = 20.66 \, \text{MPa}, \quad VR = 0.15 \quad (8.22)$$

This void ratio value represents the total porosity of the formation that includes both the matrix and natural fractures.

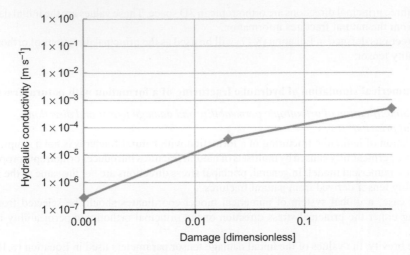

Figure 8.3   Damage dependency of permeability.

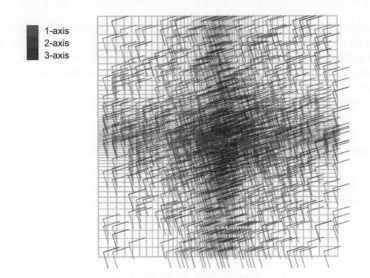

Figure 8.4   Direction of principal value of permeability tensor.

For the purpose of convenience of illustration as well as possible comparison between numerical results of damage contour and the microseismic data of a hydraulic fracturing operation, the concept of damage intensity $\bar{\omega}$ is proposed and is defined as:

$$\bar{\omega} = \sqrt{\omega_1^2 + \omega_2^2 + \omega_3^2} \tag{8.23}$$

Injection loading is applied to the central point of the model. It represents injection that occurs at an 8-m long perforation section of a vertical well, with the value of injection rate *vs.* time (Fig. 8.5). Parameter values in this diagram are provided for convenience of calculation. The negative sign in the vertical axis indicates "injection flow" other than "flow-out" rate.

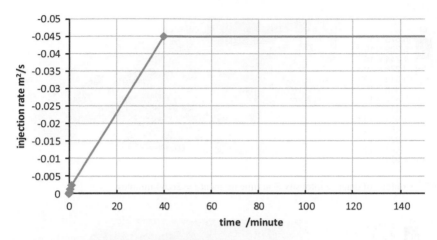

Figure 8.5   Diagram of injection rate *vs.* time.

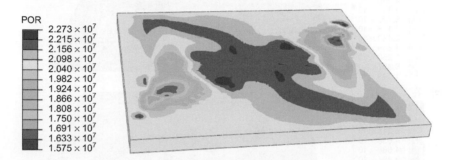

Figure 8.6   Contour of pore pressure (in Pa) at time moment $t = 26.64$ minutes during injection.

8.6.3.3   *Numerical results: principal directions of natural fracture differ from those of geostress*

Numerical simulation of the hydraulic fracturing operations was performed by finite element (FE) software using the information presented previously. As previously described, the principal directions of natural fractures are different from those of geostress. Figures 8.6 through 8.8 show the numerical solution of pore pressure distribution, damage propagation, and shear strain intensity γ_T's evolution for two time moments $t1 = 17.54$ and/or $t2 = 26.63$ (minutes) during injection loading. Figure 8.6 shows the contour of pore pressure at a time moment. Figure 8.6 shows that the pore pressure migrates along the direction of the principal permeability value at the area around injection point. However, a low pore pressure area is formed ahead of the region of higher pore pressure. Consequently in the area away from injection point at a distance of about 40 m, the path of the pore pressure migration kinks to another direction. Figure 8.7 shows the contours of damage intensity $\bar{\omega}$ at time moment $t2 = 26.63$ (minutes). The value of damage intensity $\bar{\omega}$ is the total value of continuum damage caused by the natural fracture and hydraulic fracturing operations. Figure 8.8 shows the contour of shear strain intensity γ_T, which is defined as $\gamma_T = \sqrt{\gamma_{12}^2 + \gamma_{13}^2 + \gamma_{23}^2}$ and used for shear strain intensity illustration purposes.

Figure 8.6 shows a boundary effect in this pore contour generated by hydraulic fracturing operations. The pore pressure contour that is near the boundaries is significantly nonuniform. Therefore, only the solution of P_p some distance away from the boundary should be used for analysis.

Figure 8.7 shows that the synthetic damage contour forms a complex fracture network in the near field of the injection point. The direction of major damage distribution bands along the two

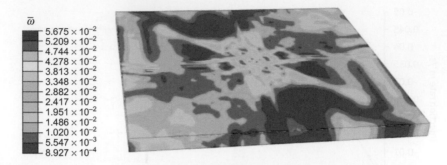

Figure 8.7   Contour of damage intensity $\bar{\omega}$ at time moment $t = 26.64$ minutes.

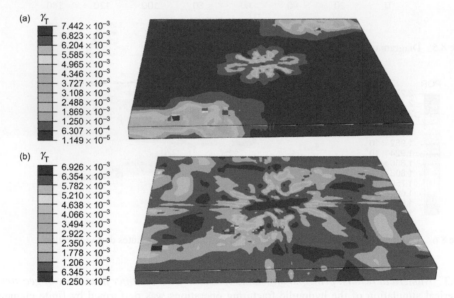

Figure 8.8   Contour of shear strain intensity γ_T at two time moments: (a) $t = 17.53$ minutes; and (b) $t = 26.64$ minutes.

principal directions of damage tensor, which is actually the directions of natural fractures. This phenomenon is of important significance to the work on optimized fracturing design.

The contour evolution of shear strain intensity γ_T's shown in Figure 8.8a and 8.8b indicates the location of expected micro seismic activity, which can be recorded by monitoring the shear sonic data during a hydraulic fracturing operation.

8.6.3.4   *Numerical results: principal directions of natural fracture overlap with those of geostress*

For the purpose of comparison, an example of hydraulic fracturing is presented for the case that no angle exists between the directions of principal stresses and that of natural fractures.

The values of all other data are the same as those used in Section 8.6.3.2.

In this case, there is no angle between NF and that of geostress. Figure 8.9 through Figure 8.11 show the numerical results. Figure 8.9 shows that the pore pressure contour develops in directions of two horizontal principal stress components, and develops in a symmetrical pattern.

Figure 8.10 shows the fracture network generated by hydraulic injection. This phenomenon of networking of the damage contour actually is generated by damage localization. The mechanism

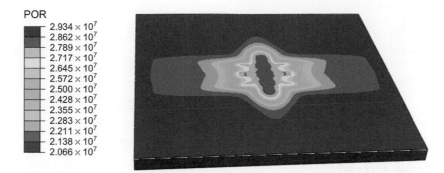

Figure 8.9   Contour of pore pressure [Pa] at time moment $t = 17.53$ minutes.

Figure 8.10   Contour of damage intensity $\bar{\omega}$ at time moment $t = 17.53$ minutes.

Figure 8.11   Contour of shear strain intensity γ_T at time moment $t = 17.53$ minutes.

behind this phenomenon could be plural. One of the mechanisms should be the material instability related to strain-softening. The other factor is relevant to the porous behavior that pore pressure takes the path along which the permeability value is higher than that of other places. The interaction of these two mechanisms intensifies the phenomenon of fracture networking generated by injection.

Because the natural fracture is homogenized as continuum damage and embodied into permeability, induced fracture networking is related to natural fractures, but does not directly come from natural fractures. There is no network of natural fractures in the model presented here.

Figure 8.11 shows contour of shear strain intensity γ_T. It is seen that the value of γ_T is smaller in the very center of the model where there is higher value of damage intensity as shown in Figure 8.10. Boundary effects are not significant for these sets of numerical suctions with the

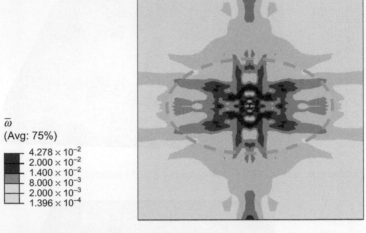

$\bar{\omega}$
(Avg: 75%)

- 4.278 × 10^{-2}
- 2.000 × 10^{-2}
- 1.400 × 10^{-2}
- 8.000 × 10^{-3}
- 2.000 × 10^{-3}
- 1.396 × 10^{-4}

Figure 8.12   Contour of damage intensity $\bar{\omega}$ at time moment $t = 36.64$ minutes.

current scale of 200 m, but they seem to be significant in the results shown in Figure 8.6 through 8.8. For the case which the principal directions of natural fractures are not the same as that of principal stresses, its scale should be larger than 200 m.

8.6.3.5   *Remarks on the accuracy of numerical results: fractures near-wellbore and in far field*

Fracture development in the hydraulic fracturing process can be divided into two parts: near-wellbore fracture (within 15 ft (approximately 4.5 m) from the borehole) and far-field fracture (15 ft from the borehole) (Verga *et al.*, 2002; Williams *et al.*, 2016). Because of the stress concentration and other factors, such as drilling, near-wellbore fractures are rather complicated: mode-I, mode-II, and mode-II cracks are all possible, and their path of development is most likely in the form of a curve or 3D surface. A detailed description of the near-wellbore fracture, however, is not as important to the final formation stimulation result as that of the far-field fracture. The production quantity is more affected by hydraulically induced fractures in the far-field fractures, rather that the fractures near-wellbore.

Therefore, the numerical results of fracture development presented in previous sub-Sections 8.6.3.3 and 8.6.3.4 are obtained by using a model without a detailed borehole geometry description.

In the lab, fracture testing can only focus on the near-wellbore crack. This makes the lab testing of hydraulic fracturing less useful to production prediction than that of the numerical results.

To further understand the fracture development under hydraulic fracture in the far field, Figure 8.12 and Figure 8.13 provide contours of damage intensity $\bar{\omega}$ at time moment $t = 36.64$ and 46.64 minutes, respectively. Plane view is used in these two figures because the focus is the fracture development in the horizontal plane.

Because x-direction is the direction of *SH*, the fracture network formed by a localized band of damage intensity $\bar{\omega}$ develops more in the x-direction than in the y-direction. Consequently, it forms an area of fractures in the elliptical form, as shown by the dashed curve in Figure 8.12. The dark area (which indicates a more fractured area) connected to the border in y-direction in Figure 8.12 is considered as a result of boundary effect and thus is regarded as not true. The reason of this is that a real fractured area should develop continuously from center to border.

The contour of damage intensity $\bar{\omega}$ at time moment $t = 46.64$ minutes shown in Figure 8.13 illustrates the continuous development of the fracture network formed by a localized band of damage intensity. The area of fractures in the elliptical form as the dashed curve area develops on the basis of the one shown in Figure 8.12. The dark area connected to the border in the y-direction in Figure 8.13 is considered to be a result of boundary effect and thus is regarded as not true.

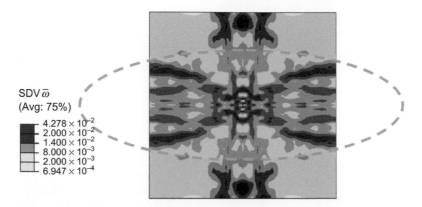

SDV $\bar{\omega}$
(Avg: 75%)

- 4.278×10^{-2}
- 2.000×10^{-2}
- 1.400×10^{-2}
- 8.000×10^{-3}
- 2.000×10^{-3}
- 6.947×10^{-4}

Figure 8.13   Contour of damage intensity $\bar{\omega}$ at time moment $t = 46.64$ minutes.

8.7   CONCLUSIONS

This chapter introduces the process for modeling hydraulic fracturing operations in reservoirs with multisets of natural fractures using the continuum-damage method. It also shows that the coupled hydro-mechanical FEM is an effective tool for estimating stimulated volume in tight-sand reservoirs.

There is an initial damage tensor for each set of natural fractures. The total initial damage tensor can be calculated by a simple summation of those for each single natural fracture set. Because the initial principal damage value is significantly less than 1.0, this simple summation of the damage tensor calculation is proved practical for modeling initial natural fractures.

The example presented shows a 73° angle between the direction of minimum horizontal stress and the principal direction of the damage tensor and the orthotropic permeability tensor.

The concept of damage intensity $\bar{\omega}$ is introduced to illustrate the intensity of fracturing generated by hydraulic injection. It does not have any independent mechanical meaning.

The numerical solution of pore pressure contour and damage contour generated by injection are presented. The contour of synthetic damage forms a complex fracture network in the near field of the injection point. The direction of the major damage distribution band turns from the direction of *SH* to the direction of natural fractures. This indicates that the directions of principal stresses and natural fractures should be considered and accounted for in related calculations to optimize hydraulic fracturing design.

ACKNOWLEDGEMENTS

The authors thank Guoyang Shen for contributing to the development of the user subroutine used in this chapter and other works on numerical modeling.

CHAPTER 9

Construction of complex initial stress field and stress re-orientation caused by depletion

Complex initial stress refers to the geostress field in which the orientation of maximum horizontal stress (*SH*) varies with location. In one case provided in this chapter, the orientation of *SH* in the upper formation section differs from that of the lower section. In another case, the orientation of *SH* on the west side of a field differs from the orientation on its east side. Special approaches to construct the initial geostress field that matches phenomena observed at the well locations are necessary in these cases.

This chapter introduces the principle and methods used to analyze *SH* rotation and initial geostress direction. Equations and user subroutines used to construct these types of complex initial stress fields are presented. Two examples are also presented to illustrate the workflow.

9.1 INTRODUCTION

The direction of initial stress field is an important factor that helps to determine the success of a hydraulic fracturing operation design. An induced fracture can be generated in the direction normal to the trajectory of a horizontal well section if its axial direction is along the direction of minimum horizontal stress (*Sh*), which consequently results in the maximum possibility of gas/oil production. Otherwise, if the direction of the well trajectory is not in the direction of *Sh*, the generated fracture will propagate in an unfavorable direction. This results in significantly less gas/oil production. Therefore, analysis of initial geostress is the first key step during the design of hydraulic fracturing operations and other related works.

The initial stress field balances the load of gravity. In principle, if the formation tops are flat, and there is no salt body and/or fault, the process to determine the directions of initial geostress principal components is simple. The values of vertical stress can be obtained by the integration of gravity along the vertical depth, and values of horizontal stress components can be calculated using the elastic theory in terms of Poisson's ratio.

However, if the history of regional tectonic movement is strong, the process to determine the directions of initial geostress principal components is complex. The following sections describe a specific case along the well trajectory in which the directions of principal stresses in the upper part of the formations can be 90° different from those in the lower part.

In general, several tectonic and depositional mechanisms influence the orientations of principal stresses:

- Accumulation of stresses adjacent to faults before slippage.
- Halo kinetics (i.e., movement of salt masses).
- Relaxation of stresses adjacent to faults.
- Slumping.
- Rapid deposition of sediments on top of a subsurface environment dominated by strike-slip or reverse faulting.

Martin and Chandler (1993) reported a *SH* rotation near two major thrust faults that were intersected during the excavation of the Underground Research Laboratory (URL) in the Canadian Shield. The fault system divides the rock mass into varying stress domains in this region. Above the fault system, the rock mass contains regular joint sets in which the *SH* is oriented parallel to the major subvertical joint set. Below the fault system, the rock is massive with no jointing; the *SH* has rotated approximately 90° and is aligned with the dip direction of the fracture zone. Stress rotation is commonly observed where the block of rock above the fault has lost its original load because of displacement above the fault. This results in considerably less *SH* magnitude than the magnitudes below the fracture zone in which the *SH* magnitude is fairly constant (Martin and Chandler, 1993). The *Sh* also increases near the faults and decreases above and below the fault zone. Teufel *et al.* (1984) observed stress rotation with depth at the multiwell experiment site in Rifle, Colorado, USA from N75°W in the upper fluvial zone (at 1501 m) to N89°W in the coastal zone (at 1980 m) to N75°W in the marine zone (at 2410 m). This observation is a result of the stress associated with large local topographical loading superimposed on the regional stress field of the basin.

Miskimins *et al.* (2001) observed stress rotation in a complex faulting field (North LaBarge Shallow Unit of the Green River basin), which is located in a major thrust fault, a tear fault, four major strike-slip faults, and a "horse-tail" splay termination. The regional *SH* orientation is consistent in the north-south direction. However, the wells associated with the thrust fault and major strike-slip fault exhibit stress rotation between 45 and 90° because of the proximity to the major strike-slip and reverse fault. This is likely attributable to changes in tectonic stresses throughout geological time.

Depletion can also induce stress rotation and impose high pressure on casing by direct compaction of the reservoir rock, overburden fault, and bedding plane movement triggered by the reservoir compaction and in-well mechanical hot spots. Kristiansen (2004) discussed the compaction of the chalk formation in the Valhall field, which caused a seafloor subsidence of nearly 5 m and stress reorientation of approximately 90°. The subsidence continues at 0.25 cm/year, and a major effort has been expended to understand the significant effect of wellbore stability, which creates the surface changes. The surface changes are followed by depletion, compaction, and subsidence, which also reactivate faults and shear casing.

The perturbation of stresses adjacent to salt bodies is significant and depends on the geometry of the salt. Salt affects the actual geomechanical environment through the alteration of the local stresses because salt cannot sustain deviatoric stress and deforms by means of plastic creep in response to any imposed deviatoric stress. Consequently, changes in the stress field and shear stresses adjacent to the salt are sufficient to cause a reorientation of the principal stress. It is well known that a rubble zone occurs below or adjacent to the salt diapirs, which is an intrinsic consequence of the equilibrium stress field necessary to satisfy the various stresses that exist within the salt body and the adjacent formation.

This work presents a numerically constructed initial stress field with a case of complex stress distribution for both a local model with only one well and a block model with multiple wells. Details of the numerical model aspects are introduced.

Numerically simulated stress reorientation caused by pore pressure depletion and liquid injection is also presented.

9.2 CONSTRUCT INITIAL STRESS FIELD WITH A LOCAL MODEL OF COMPLEX STRESS PATTERN

This section presents numerical modeling of initial geostress with complex stress pattern in tight-gas reservoir formations. *SH* orientation varies with a true vertical depth (*TVD*) from NW30° in the upper part of the formations to NE60° in the lower part of the formations.

9.2.1 Geology and one-dimensional (1D) geomechanics solution

Figure 9.1 shows the regional geology of the field. The formation top of the reservoir is shown on the left and the sectional view of the geostructure near the target well to be fractured is shown

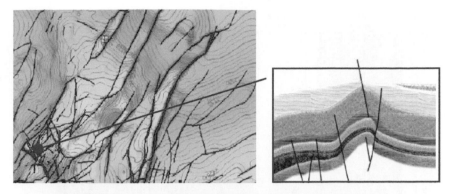

Figure 9.1   Geology of the region where the target well is located (after Guenifi, 2013).

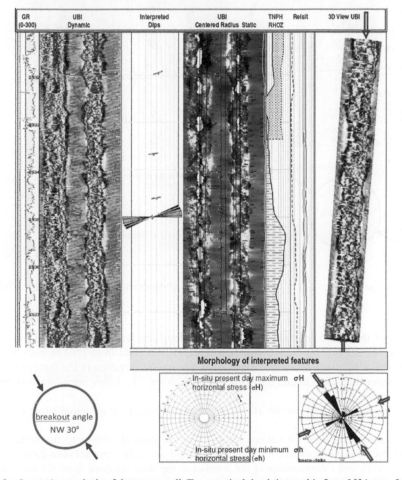

Figure 9.2   Image log analysis of the target well. True vertical depth interval is from 2531 m to 2538 m.

on the right. The geological history of tectonic movement in this area is strong. Several faults exist and an anticline structure is identified.

An analysis of the image log and breakout angle along the target well shown in Figure 9.2 indicates that the *SH* is in the NW30 direction at the *TVD* interval from 2531 to 2538 m.

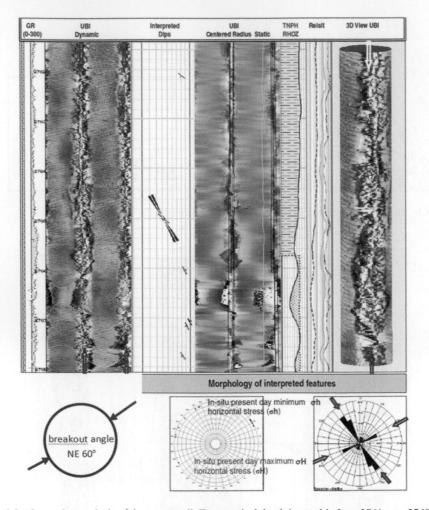

Figure 9.3   Image log analysis of the target well: True vertical depth interval is from 2761 m to 2768 m.

An analysis of the image log and breakout angle along the target well shown in Figure 9.3 indicates that the *SH* is in the NE60° direction at the *TVD* interval from 2761 to 2768 m.

In summary, a 90° angle difference exists between the *SH* direction in the formation section at a *TVD* greater than 2500 m and a *TVD* less than 2700 m. The following subsection provides additional details.

Table 9.1 shows the results of the 1D geomechanical analysis of principal stress component values.

There are a total of eight formations layers from the ground surface to the bottom of the model. This includes two reservoir formation layers and one tight-zone between them. Table 9.1 shows the density and z-coordinates of the formation top data obtained from the measurement of two wells. The density value is the same for the same type of formation, even though is some difference in the z-coordinates of their formation top. The location of the formation tops is given in terms of z-coordinates. The relationship between the z-coordinates and the formation top *TVD* is z-coord $= 3005 - TVD$ [m].

The effective stress ratio is $k0 = 0.5$, and the tectonic factor $= 0.875$ for all of the formation layers. The pore pressure value is equal to the hydrostatic level.

Table 9.1   Values of principal stress components obtained using 1D geomechanical analysis.

Well 1

	S_v, Well 1 [Pa]	z-coordinate of Formation top [m]	ρ [kg/m^3]
Layer 1	0.00	3.05×10^3	2.40×10^3
Layer 2	7.52×10^6	2.74×10^3	2.50×10^3
Layer 3	1.47×10^7	2.44×10^3	2.63×10^3
Layer 4	2.40×10^7	2.08×10^3	2.46×10^3
Reservoir 1	3.99×10^7	1.42×10^3	2.63×10^3
Target zone (TZ)	5.15×10^7	9.70×10^2	2.68×10^3
Reservoir 2	5.37×10^7	8.86×10^2	2.64×10^3
Bottom	6.95×10^7	2.75×10^2	2.67×10^3

An accurate approach to construct the three-dimensional (3D) initial geostress is to build a numerical model using the exact geostructure, which comprises the depth variation of the formation tops as well as faults and anticline structures (Fig. 9.1). A significant amount of seismic data and geological analysis as well as efforts to convey the geometry in a finite element (FE) mesh or finite difference (FD) grid are necessary when building this type of model.

A simple approach to construct the initial geostress field in practice can be realized through a local numerical model that does not honor the geostructure. The model only honors the stress solution obtained from 1D geomechanical analysis with the image log and breakout angle of the target well (or a set of offset wells in general).

9.2.2 Finite element model

Figure 9.4 and 9.5 show the FEM meshes. The submodeling technique is a specific method for bridging the gaps of scales comprised in a 3D geomechanics problem (Dassault Systèmes, 2008). In this technique, the submodel is a small portion of the global model, which represents the entire structure. Precision with submodel analysis can be significantly improved by refining the mesh of the submodel. The global model is shown on the left of Figure 9.4, and the submodel is shown on the right. The global model is 12,000 m wide, 15,000 m long, and 3005 m thick. The submodel is 10 m wide, 10 m long and 95 m thick. The model sizes presented here are for workflow illustration purposes. In engineering practice, the submodel size should be as large as necessary.

Function of submodel shown in Figure 9.5 is to reproduce the phenomena of break-out angles as that shown in Figures 9.2 and 9.3. In this way, it is sure that the constructed stress field reproduces the phenomenon observed with logging data.

The global model and submodel are adopted to apply a nonzero displacement boundary condition, thus avoiding distortion in the areas near the bounds of the model.

Boundary conditions of the global model are zero normal displacement constraints to all four lateral surfaces plus the bottom. The top surface is free, which represents the ground surface. Boundary conditions of the submodel are nonzero displacements in all three directions and all surfaces. Nonzero displacement values at these surfaces are subtracted from a numerical solution of displacement in the global model. Drilling mud pressure is applied to the inner surface of wellbore as shown in Figure 9.5.

The task of assigning initial geostress values in reference to the 1D geomechanical solution can be challenging. Principal stress rotation should be performed while assigning the stress value to the FEM elements because of the variation of principal stress directions within formations at 2500 to 2700 m *TVD*. This can be accomplished by using a user subroutine that is specially developed for this purpose.

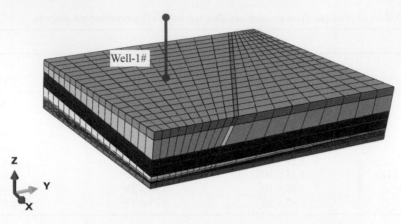

Figure 9.4   Mesh, layers, and geometry of the global model.

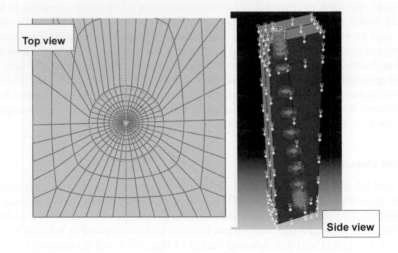

Figure 9.5   Illustration of mesh and geometry of submodel.

The following assumptions are adopted in the calculations:

- *X* and *Y* directions of the numerical model are east and north, respectively.
- The original vertical stress is one of the three principal stresses, and no rotation is performed in relation to the vertical axis. Rotations are only performed on the horizontal stress components.

The following content is the source code written in FORTRAN programming language for the rotation of principal stress direction in terms of azimuth direction and the coordinate system of the numerical model. Explanations are inserted into the code with lines that begin with the letter 'c'.

```
c-------start of source code---------------------
c Major principal direction NW 30 (SE30) from depth 0 to 2565 m
c That corresponds to z-coordinate 3060 to 442 m
c Original direction of maximum horizontal stress sigCH is
    assumed to be in Y-direction,
c minimum horizontal stress sigmash is in X-direction.
```

```
c A ngle-1=30, angle-2=120 degree. Here it calculates the
  directional cosines.
   pi=3.1415926535
   tpi=pi/180.
   t11=cos(30.*tpi) !l1 x x1
   t12=cos(60.*tpi) !m1 y x1
   t21=cos(120.*tpi) !l2 x y1
   t22=cos(30.*tpi) !m2 y y1
c
c Secondary principal directions where changes 90 degree from
  above NE 60/SW60
c from depth 2638 to bottom angle-1=120, angle-2=210 degree.
  Here it calculates the
c directional cosines after 90 degree rotation for bottom
  formations.
   ct11=cos(120.*tpi) ! x x2
   ct12=cos(30.*tpi) ! y x2
   ct22=cos(210.*tpi) ! x y2
   ct21=cos(120.*tpi) ! y y2
c
c He re it calculates the principal stress Sv (sigma0),
  SH=0.875*Sv, Sh=0.5*Sv
c Effective stress is used here.
c
if(z.gt.2612) then
   sigma0=-4.8e6
   sigma(3)=(3060-z)/(3060-2612)*sigma0
   sigmach=sigma(3)*0.875!sigmay
   sigmash=sigma(3)*0.5 !sigmax
c
c Here is calculates stress rotation for points of given TVD
  depth and directional angle
c sig-x1=l1**2*sig-x+m1**2*sig-y
c
   sigma(1)=t11**2.*sigmash+t12**2.*sigmach
c
c sig-y1=l2**2*sig-x+m2**2*sig-y
c
   s igma(2)=t21**2*sigmash+t22**2.*sigmach
c
c tao-xy1=l1*l2*sig-x+m1*m2*sig-y
c
   sigma(4)=t11*t21*sigmash+t12*t22*sigmach
   sig ma(5)=0.
   sigma(6)=0.
c---end- - - - - - - - - - - - - - - - - - - - - -
```

The *SH* direction rotates gradually from NW30 to NE60° for the *TVD* interval. Thus, the *SH* direction angle at a *TVD* point varies with its value of *TVD*. This variation should be considered during the calculation of stress rotation.

The stress of each layer with different *TVD* values is assigned to the elements of the numerical model by repeating the stress calculation process.

This set of user subroutine is used to assign initial geostress for both global model and submodel.

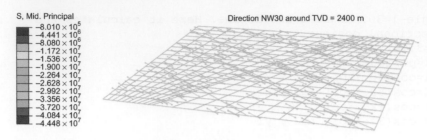

Figure 9.6   Direction of *SH* at *TVD* = 2400 m: NW30°.

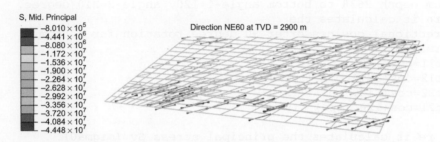

Figure 9.7   Direction of *SH* at *TVD* = 2900 m: NE60°.

Figure 9.8   Direction of *SH* varies from NW30 to NE60° at the *TVD* interval at 2600 m.

9.2.3   Numerical results

Figure 9.6 shows the stress contour of the global model generated using the user subroutine defined for stress rotation at *TVD* of 2400 m (*z*-coord = 650 m). Figure 9.7 shows the stress contour at *TVD* of 2900 m (*z*-coord = 150 m). The direction of *SH* at *TVD* = 2400 m is NW30 and NE60° at *TVD* = 2900 m.

Figure 9.5 through Figure 9.8 show *SH* as the middle principal stress component. The variation of *SH* direction at this *TVD* interval is not uniform.

Figure 9.9 shows the situation of the direction variation of *Sh*, which is regarded as the maximum principal stress with sign convention of solid mechanics. Figure 9.9 shows the direction of *Sh* varies gradually from NE60° at 2550 m to NW30° at 2630 m.

Figure 9.9 shows the stress contour of the global model generated using the user subroutine defined for stress rotation at *TVD* of 2400 m (*z*-coord = 650 m). Figure 9.6 shows the stress contour at *TVD* of 2900 m (*z*-coord = 150 m). The direction of *SH* at *TVD* = 2400 m is NW30 and NE60° at *TVD* = 2900 m.

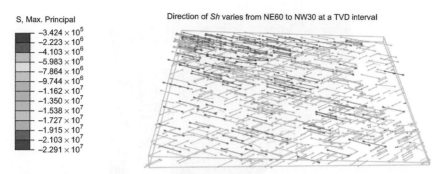

Figure 9.9 Direction of *Sh* varies from NE60° (light green, 2550 m) to NW30° (dark blue, 2630 m) at the *TVD* interval of 2550 to 2630 m.

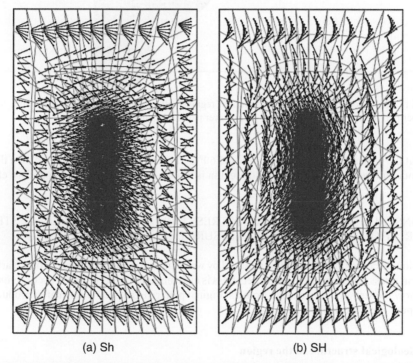

(a) Sh   (b) SH

Figure 9.10 Top views of numerical solution of submodel: (a) Direction of *Sh* varies from NE60° (2550 m) to NW30° (2630 m) within the *TVD* interval of 95 m. (b) Direction of SH varies from NW30° to NE60°.

Figure 9.6 through Figure 9.8 show *SH* as the middle principal stress component. The variation of *SH* direction at this *TVD* interval is not uniform.

Figure 9.10 shows numerical solution obtained with submodel shown in Figure 9.5. The two tops shown in Figure 9.10 indicate that (i) direction of minimum horizontal stress *Sh* varies from NE60° (2550 m) to NW30° (2630 m) within the *TVD* interval of 95 m, and (ii) simulataniously direction of SH varies from NW30° to NE60°.

Variation of break-out around the borehole is shown in Figure 9.11 with a cut-side-view of plastic region of the submodel. The red-colored mesh is the plastic region at where plastic deformation occurs with tensile failure, and this is the break-out described with principle of geomechanics. Cut-side-view of numerical solution of plastic regions shows that break-out angle varies from

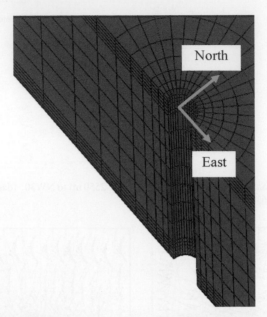

Figure 9.11　Cut-side-view of numerical solution of regions with plastic deformation: varies from NW30° at the upper part around borehole to NE60° at the lower part of the submodel.

NW30° at the upper part around borehole to NE60° at the lower part of the submodel. On the top of the model, numerical solution of plastic region is too much to be true due to boundary effect.

9.3　CONSTRUCTION OF INITIAL GEOSTRESS FIELD AND SIMULATION OF STRESS VARIATION CAUSED BY PORE PRESSURE DEPLETION

This section discusses the stress rotation caused by water injection and/or oil production. Variation of the local stress direction of the initial geostress is also modeled. A 3D model is used. Numerical results show the stress pattern change from its original pattern to a specific form after production-induced pore pressure depletion.

9.3.1　Geological structure in the region

Figure 9.12 and Figure 9.13 show the partial cross-sectional view of the KS field geostructure and the plane view of the reservoir formation top. Figure 9.13 shows that the *SH* orientation varies around the circumferential direction of the formation top where a salt body exists. Measured data of the *SH* orientation is NN0° at the Well 1 location and NE80° at the Well 2 location (Zhang *et al.*, 2015).

9.3.2　Gas production plan

Figure 9.14 shows the 2-year gas production plan for Well 1. The objective of the numerical calculation is to predict the variation of the *SH* direction after production during this period.

　　To achieve this objective, the following two numerical modeling tasks are performed:

- Construct the initial 3D geostress using a 1D solution geomechanical analysis, which includes three principal geostress components and pore pressure.

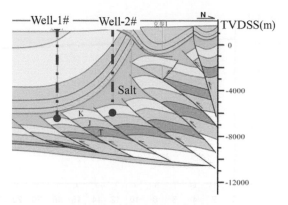

Figure 9.12   Cross-sectional view of the KS field geostructure (after Zhang *et al.*, 2015).

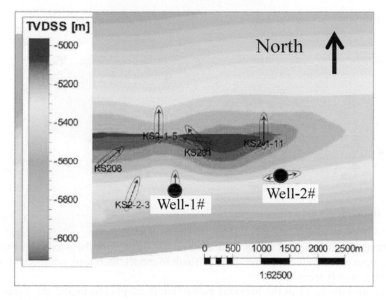

Figure 9.13   Plane view of the reservoir formation at the top of the KS field (after Zhang *et al.*, 2015).

- Perform a hydro-mechanical coupled analysis to simulate the porous flow and matrix deformation of the reservoir formations, thus obtaining the numerical solution of stress contour of the depleted field.

For workflow illustration purposes, the simplified approximate model is used in the calculation. There is no salt body specification in the simplified model; only the phenomena of variation of stress orientation and magnitude of stress are modeled. This approach is practical for the prediction of stress variation *vs.* pore pressure depletion with reasonable accuracy for a specific time period. However, it should be noted that the solution of stress variation is more accurate when an exact salt body specification is included.

9.3.3   Finite element model

Figure 9.15 shows the FE mesh of the numerical model for this task. The *y*-axis is in the north direction, and the *x*-axis is in the east direction. The *SH* component direction at Well 1 overlaps the *y*-axis. Therefore, initial geostress input for the data obtained at the Well 1 location is easy to

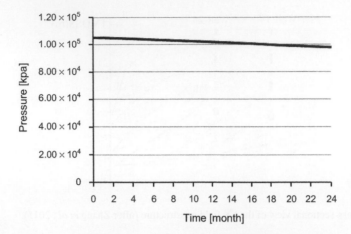

Figure 9.14   Pore pressure depletion derived from the 2-year gas production plan for Well 1.

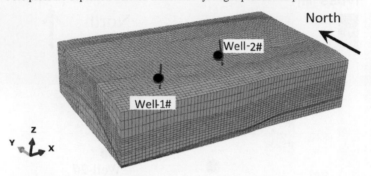

Figure 9.15   Illustration of the well locations.

model. However, the direction of *SH* at Well 2 has an 81° angle with the *y*-axis. This makes the initial stress input at the Well 2 location more difficult to model; therefore, additional rotation of the stress orientation is necessary.

The procedure introduced in the previous subsection is adopted to model the stress rotation at the location of Well 2 while importing the stress orientation data into the 3D FEM shown in Figure 9.15.

9.3.4   Numerical results

Figure 9.16 shows the resultant *SH* distribution of the initial geostress by omitting the calculation detail in the model. For brevity, only the formation layer direction is shown. Its depth is shown in the middle of the model. The direction of *SH* is NN0° at the Well 1 location and NE81° at the Well 2 location.

Figure 9.17 shows the *SH* distribution after local stress equilibration between the assigned value and the one generated by elastic calculation using the model formation structures. The *SH* direction at Well 2 slightly changes from its original uniform direction of 81° and becomes a stream-lined curve around the well.

Figure 9.18 and Figure 9.19 show the numerical results of the *SH* directions after pore pressure depletion. Figure 9.18 shows the *SH* direction distribution after depletion in the XoY horizontal plane. Figure 9.19 shows the *SH* direction distribution after depletion in the XoZ vertical plane.

Figure 9.18 shows that the *SH* direction varies from its original direction before pore pressure depletion. This is attributed to the stress disturbance caused by the subsidence of the upper

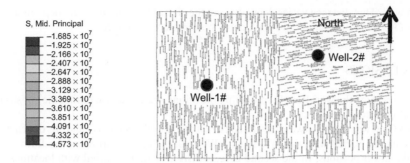

Figure 9.16   Resultant *SH* distribution of the initial geostress assigned to the model.

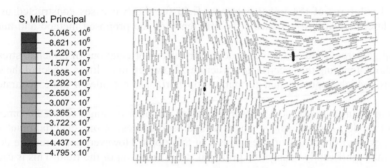

Figure 9.17   Resultant *SH* distribution of the initial geostress equilibrated with geostructures.

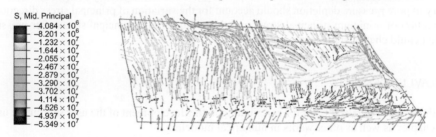

Figure 9.18   Resultant *SH* distribution after pore pressure depletion: XoY plane view, *TVD* is located in the upper part of reservoir formation.

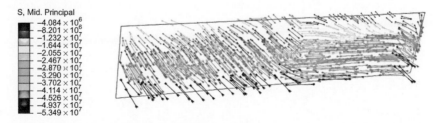

Figure 9.19   Resultant *SH* distribution after pore pressure depletion: XoZ vertical cross-sectional view.

formation. Figure 9.19 shows that the *SH* direction in the lower section of the model does not vary as much as the upper section of the model.

Figure 9.19 also shows that the stress reorientation phenomena after pore pressure depletion occurs near the production well. There is no significant stress reorientation far away from the production well. This distance varies depending on the size of the pore pressure depletion area as well as the *TVD* of the formation in which pore pressure depletion occurs.

9.4   CONCLUSIONS

This chapter presents the workflow for modeling the initial stress rotation during the construction of the initial geostress field.

This workflow was used in two cases to construct the initial geostress field. In the first case, the *SH* orientation varied with the *TVD* from NW30° in the upper portion of the formations to NE60° in the lower portions. In the second case, the *SH* orientation varied with locations in the horizontal plane from NN0° in the left section of the model to NE81° in the lower-right section (Fig. 9.13).

Numerical solutions of the initial geostress field for the two cases constructed using the previously described workflow indicate the efficiency of the proposed method and numerical model.

The initial geostress field can be constructed for various orientations of principal stress components without honoring the specific geostructures. However, in reality, a model of the specific geostructures, which honors actual formation top horizons, provides a more accurate stress solution.

Stress orientation caused by pore pressure depletion is modeled using a coupled geomechanics and reservoir model. Theoretically, changes of hydrostatic stress at one material point will not change the principal directions of stress tensor at this point. However, pore pressure depletion of a region can significantly change the principal stress direction of the field. This is caused by nonuniform deformation of the specific geostructures.

The numerical results presented in this chapter suggest that the trajectory design in a field with a history of pore pressure depletion should account for the variation of principal stress directions. The direction of minimum principal stress and that of maximum principal stress after pressure depletion could change significantly.

ACKNOWLEDGEMENTS

The authors thank Guoyang Shen for contributing to the development of the user subroutine used in this chapter as well as other works in numerical modeling.

CHAPTER 10

Information transfer software from finite difference grid to finite element mesh

In many cases, pore pressure analyses were performed using the finite difference (FD) method, and three-dimensional (3D) geomechanics were analyzed using the finite element (FE) method. To perform a coupled analysis for porous flow and matrix deformation (such as subsidence calculation), it is necessary to transfer data between the FD grid and FE mesh. This chapter introduces a method for developing a data-transfer platform.

10.1 INTRODUCTION

Abaqus™ finite element analysis (FEA) is an effective 3D FE tool for various types of geome-chanical analysis and has been widely used in the petroleum industry in the recent decade (Capasso and Mantica, 2006; Dean *et al.*, 2003). Pore pressure and reservoir porosity are essential input data for this type of mechanical analysis and are usually the results of reservoir analysis. Reservoir analyses typically use the FD method to solve mathematical problems of porous flow within the formation, thus deriving the distribution of pore pressure. An Abaqus FE mesh typically uses a different number of nodes than the FD grid for various reasons. Therefore, a tool that enables data transfer between the FD grid and Abaqus FE mesh is necessary.

Capasso and Mantica (2006) introduced the method upon which the FE mesh was built by importing and modifying the FD grid. However, because the FE mesh domain of interest analyses are generally not the same as the FD analysis, the Capasso-Mantica method can only be used in a few cases.

Dean *et al.* (2003) presented comparison studies on the method that coupled the geomechanical model with reservoir fluid flow. The same mesh for both FE and FD analyses is used.

In this study, a software set is developed that can transfer data from the FD analysis grid to the FE mesh nodes. The FE mesh domain can be the entire FD grid or only a portion of the FD grid. Using the data transfer platform presented in this chapter, both reservoir and basin data (e.g., pore pressure and porosity) as results of a FD analysis can be transferred precisely and efficiently from the relevant FD grid to the FE mesh nodes.

This chapter presents important information that will benefit all FEM users in the petroleum industry who engage in geomechanical analyses. This includes subsidence analysis, hydraulic fracturing of reservoirs, and wellbore stability analysis in which pore pressure and porosity data are used as input data.

10.2 DESCRIPTION OF PRINCIPLE

The function of the platform is to transfer data from the FD grid to the FE mesh. The former is the resultant data from reservoir analyses, and the latter is the input data for geomechanical analyses. Figure 10.1 shows the data transfer procedure presented in this work. A source FD grid is imported along with its nodal geological data and/or mechanical data as source data. A target FE mesh is also imported as source data. The relative position of the center of the FE mesh in relation to the FD grid is specified by the input data. The software then calculates the

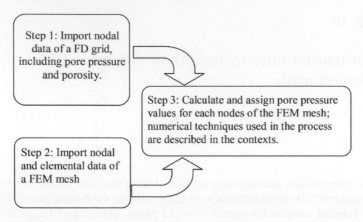

Figure 10.1　Flow chart of the data transfer procedure.

FE mesh nodal information of mechanical variables (such as pore pressure) point-by-point from the corresponding FD grid nodal information of the same mechanical variable. An optimized domain radius of each FE mesh node is obtained based on the elemental information. This local calculation domain radius is applied to the FD source grid. Therefore, only the FD grid nodes that fall within the scope of this radius are used to calculate the value of a variable at a target FE mesh node. The mechanical variable value at a target FE mesh node is calculated by using a weight factor that depends on the FE mesh geometrical information. These calculations provide more accurate FE mesh mechanical and geological variables.

A distance-weighted calculation method is adopted to perform this transfer. The following major features are introduced:

- Unlike the Capasso-Mantica method that requires the nodes in the target FE mesh be a subset of the source FD nodes, this process enables users to construct the target FE mesh independently of the source FE mesh when the FE mesh is positioned in relation to the FD mesh. This enables the user to consider additional geological and mechanical information when constructing the target FE mesh.
- Various methods can calculate the values of a variable at target FE mesh nodes, depending on the various situations, respectively. An optimized domain radius is provided here; thereby, only the nodes that fall into that FD grid radius are used. The optimized radius of the local calculation domain for each FE mesh node is calculated based on the maximum distances of each FE mesh element that contains the node. This distance is the distance between a FE mesh element node and a FD grid node within the local calculation domain. The user can overwrite this radius. If the spatial position of a target FE mesh node is the same as the source FD grid node (optimized radius of the domain is zero), then the calculation defaults to the Capasso-Mantica method.
- A distance weight factor can be assigned for the nodes within the optimized local calculation domain radius. The user can choose a power-law weight and an exponential weight that is provided by the software. Other weighting algorithms can be used to match the FE mesh geological and mechanical properties more accurately.
- This software provides users more flexibility to build a FE model (FEM) mesh than other software. A source FD grid can contain geological and mechanical values obtained from either reservoir analysis or basin analysis, which can have different scales. The FEM adopted by the user can also be used for various purposes at the field scale, reservoir scale, or casing scale. This software can perform the previously described calculations with no constraints or negative influence from scale discrepancies. The calculation can be performed effectively and precisely if the FE mesh is positioned in relation to the FD mesh.

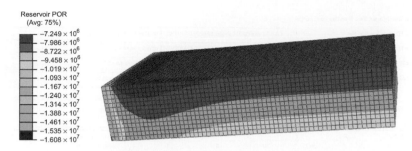

Figure 10.2   Pore pressure distribution as result of reservoir analysis along its FD grid.

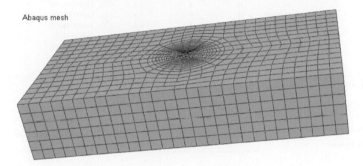

Figure 10.3   FE mesh used during Abaqus geomechanical analysis.

The spatial domain is calculated for each FE mesh node location. The spatial domain of an FE mesh node location is calculated using the domain radius of the node location. The FE mesh values of a node location are generated by identifying the FD grid node locations within the spatial domain around the FE mesh node location. Each FD grid value within the spatial domain of a given FE mesh node location is weighted based on a distance between the FD grid node location and the FE mesh node location. The FE mesh value of the node location is calculated by averaging the distance-weighted FD grid values. The resultant FE mesh node values are used during geomechanical modeling to analyze production and/or fracturing activities.

10.3   NUMERICAL VALIDATION

Figure 10.2 shows the pore pressure distributed in a field of 10 km depth, 20 km width, and 30 km length, as result of reservoir analysis along its FD grid.

Figure 10.3 shows the FE mesh that is used during geomechanical analysis. The wellbore top is located at the center of the mesh. The mesh near the wellbore has a higher density than the other parts of the mesh because the geomechanical analysis focuses on the area surrounding the wellbore. Figure 10.4 shows the interaction window of the data transfer platform that imports the FD grid spatial location and the reservoir analysis data along with the FE mesh, which is the target of the data transfer.

Figure 10.5 shows the results of the data transfer performed using the data platform. The transferred data distribution spectrum is near the original distribution shown in Figure 10.2. The transferred data deviation from its original location is less than 1%. This indicates that the data transfer platform is accurate for practical use in geomechanical analysis.

Although the geometry of the target FE mesh used here has the same geometric scale as the FD grid, the sizes of the two geometries can be different from one another. In this case, the FE mesh location should be carefully selected, based on practical engineering. The wellbore top analysis results can be used as reference data to locate the two geometries.

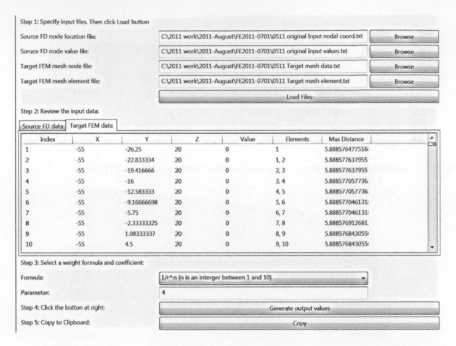

Figure 10.4   Interaction window of the data transfer platform.

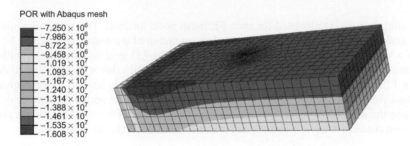

Figure 10.5   Result of the data transfer performed using the data platform.

10.4   CONCLUSION

A data transfer platform set is developed to transfer data from the reservoir analysis FD grid to the FE mesh used during geomechanical analysis. The numerical validation indicates that the accuracy level of the results using the software developed exceeds 99%. The proposed numerical procedure and the software developed provide a set of effective tools for users of various FEM methods (such as Abaqus FEM) in the petroleum industry. The procedure presented in this chapter provides a wider range of application and greater efficiency and accuracy than other tools presented in previous works of other researchers.

ACKNOWLEDGEMENTS

The authors thank Dr. Xiaomin Hu for his contribution to the development of the software presented in this chapter.

Nomenclature

A and B = loading factors, with subscript t for tension and c for compression, dimensionless

a and b = model parameters, F/L^2 [Pa]

a_1, a_2, a_3 = model parameters, dimensionless

A_f = area of the fault surface, L^2 [m^2]

Bl = brittleness index, dimensionless

Btmh = bottom hole pressure, F/L^2 [Pa]

b = model constant, dimensionless

c = cohesive strength, F/L^2 [Pa]

c_t and c_b = leak-off coefficients, dimensionless

c_β = cohesive strength defined on surfaces of natural fractures, F/L^2 [Pa]

D = the synthetic damage variable, dimensionless

DV = continuum damage variable, dimensionless

d_c = compressive damage variable, dimensionless

d_t = tensile damage variable, dimensionless

$\mathbf{E}_0^e$ = matrix stiffness of the intact material, F/L^2 [Pa]

E = young's modulus of the damaged material, F/L^2 [Pa]

E_0 = young's modulus of the initial intact material, F/L^2 [Pa]

E_{ijkl}^0 = elasticity tensor of the intact material, F/L^2 [Pa]

E_{ijkl}^{epd} = the algorithmic tangential stiffness tensor, F/L^2 [Pa]

f = yielding criterion, F/L^2 [Pa]

$\tilde{f}$ = the plastic damage loading condition, F/L^2 [Pa]

$f(\psi)$ = general form of function, dimensionless

f_t = function of damage initiation, dimensionless

F = the plastic damage potential function

F_p = plastic potential of a stress point and is defined by Equation (5.2)

F_{other} = the F_β point values along trajectories other than the optimized one, F/L^2 [Pa]

$F_{\text{optimized}}$ = F_β point values on the optimized trajectory, F/L^2 [Pa]

F_β = the CSF potential defined on the surface of natural fracture with inclination angle β, F/L^2 [Pa]

G = shear modulus of material, F/L^2 [Pa]

$G1$ = the values of fracture energy for mode-I fracture, F/L [N/m]

$G2$ = the values of fracture energy for mode-II fracture, F/L [N/m]

$G3$ = the values of fracture energy for mode-III fracture, F/L [N/m]

G_{shear} = fracture energy for mixed-mode shear crack, F/L [N/m]

G_I = fracture energy for mode-I crack, F/L [N/m]

G_{IC} = critical value of fracture G_I, F/L [N/m]

G_{II} = fracture energy for mode-II crack, F/L [N/m]

G_{IIC} = critical value of fracture G_{II}, F/L [N/m]

G_{III} = fracture energy for mode-III crack, F/L [N/m]

G_p = plastic potential function, F/L^2 [Pa]

G_T = total fracture energy for mixed mode crack F/L [N/m]

G_{TC} = critical value of total fracture energy for mixed mode crack, F/L [N/m]

$\mathbf{i}, \mathbf{j}, \mathbf{k}$ = vectors of coordinate basis, dimensionless

I_ε = first invariant of the strain tensor, dimensionless

$\bar{I}_1$ = sum of the effective principal stresses, F/L^2 [Pa]

$\bar{J}_2$ = second invariant of the deviatoric effective stress tensor, F/L^2 [Pa]

J_ε = second invariant of the deviatoric strain tensor, dimensionless

$k0$ = effective stress ratio, dimensionless

k = initial shear-strength constant, F/L^2 [Pa]

k_c = ratio between q(TM) and q(CM)

k_∞ = strain hardening limit of the fictitious net material, F/L^2 [Pa]

k_t = tangential permeability/hydraulic conductivity [m/s]

K = the ratio of a/a_t

K_i = permeability in i-th component of orthotropic principal direction [1/m^2]

K_f = fluid consistency, dimensionless

K_g = gap flow conductivity, [m/s]

K-bottom = fluid leak-off coefficient at bottom of the cohesive crack, [m/s]

k-top = fluid leak-off coefficient at top of the cohesive crack, [m/s]

K_n = interfacial stiffness in the direction normal to fracture surface, F/L^2 [Pa]

K_{t1}, K_{t2} = interfacial stiffness in the two directions tangential to fracture surface, F/L^2 [Pa]

l, m, n = directional components, dimensionless

M = magnitude of seismicity

M_0 = energy related to fault reactivation [Nm]

N = the number of sets of natural fractures, dimensionless

N_{max} = strength of material in normal direction, F/L^2 [Pa]

N_s = number of stages, dimensionless

P_p = pore pressure, F/L^2 [Pa]

p_0 = formation pore pressure value, F/L^2 [Pa]

p_t and p_b = the pore pressure on the top and bottom surfaces, F/L^2 [Pa]

p_i = midface pressure, F/L^2 [Pa]

q_t and q_b = flow rates into the top and bottom surfaces [m^3/s]

$\mathbf{q}$ = the volume flow-rate density vector [m^3/s]

q(TM) = the second invariant of the stress tensor at the tensile meridian, F/L^2 [Pa]

q(CM) = the second invariant at the compressive meridian, F/L^2 [Pa]

$\tilde{Q}$ = plastic part of the potential in the effective stress space, F/L^2 [Pa]

r = the principal stress ratio (*PSR*), dimensionless

R = stress triaxiality, dimensionless

$S1$ = the strength value in the normal direction of the fracture surface, F/L^2 [Pa]

$S2$ and $S3$ = the strength value in the two directions tangential to the fracture surface, F/L^2 [Pa]

$\tilde{s}_{ij}$ = deviatoric stress tensor, F/L^2 [Pa]

s_t = coefficient of stress state under tension, dimensionless

s_c = coefficient of stress state under compression, dimensionless

s, S = material parameters, dimensionless

S_{max} = maximum principal effective stress, F/L^2 [Pa]

S_{min} = minimum principal effective stress, F/L^2 [Pa]

Sh = minimum horizontal stress, F/L^2 [Pa]

SH = maximum horizontal stress, F/L^2 [Pa]

S_v = vertical stress component, F/L^2 [Pa]

S_h = minimum horizontal stress component, F/L^2 [Pa]

S_H = maximum horizontal stress component, F/L^2 [Pa]

T_{max}, S_{max} = strength of material in two tangential directions F/L^2 [Pa]

t_{curr}, t_{orig} = the current and original cohesive element geometrical thicknesses [m]

tf = tectonic stress factor, dimensionless

VR = void ratio, dimensionless

w_t = weight parameters that controls stiffness recovery for unloading in tension, dimensionless

w_c = weight parameters that controls stiffness recovery for unloading in compression, dimensionless

X = average value of relative displacement across the fault [m^2]

Y = thermodynamic conjugate force of damage variable d, dimensionless

Y_0 = the initial threshold of the damage-conjugate force, dimensionless

Y_t = thermodynamic conjugate force y of damage variable d for the case of loading in tension, dimensionless

Y_c = thermodynamic conjugate force y of damage variable d for the case of loading in compression, dimensionless

Z = the gap opening of a cohesive element [m]

Greek symbols

α = power-law coefficient, dimensionless

α_0 = empirical parameter for a regional's seismicity calculation, dimensionless

α_F = material constant designed for pressure-sensitivity properties, dimensionless

α_Q = dilatancy constant for the non-associated flow rule, dimensionless

α_{sf} = angle between maximum horizontal stress and direction of natural fracture [°]

β = inclination angle [°]

δ = fracture opening, L, [m]

ϕ = material parameters, dimensionless

μ_f = the internal frictional coefficient, dimensionless

$\mu_{f\beta}$ = the internal frictional coefficient defined on surfaces of natural fractures, dimensionless

μ = fluid viscosity [Ns/m]

θ = azimuth angle [°]

γ^* = parameter of triaxial stress state with Heaviside function, dimensionless

$\dot{\gamma}$ = shear-strain rate, F/L^2 [Pa]

γ = shear strain component, dimensionless

γ_T = shear strain intensity, dimensionless

δ_{ij} = second-order unit tensor, dimensionless

$\varepsilon(\theta, f_i)$ = the eccentricity of the loading surface in the effective stress space

$\boldsymbol{\varepsilon}$ = strain tensor, dimensionless

ε_t = equivalent strain variable for tensile strain, dimensionless

ε_c = equivalent strain variable for compressive strain, dimensionless

ε_{t0} = thresholds for damage initiation during loading in tension, dimensionless

ε_{c0} = thresholds for damage initiation during loading in compression, dimensionless

$\boldsymbol{\varepsilon}_{ij}$ = strain tensor, dimensionless

$\Delta\varepsilon_{ij}$ = strain increment, dimensionless

$\bar{\varepsilon}_c^p$ = equivalent plastic strain in compression, dimensionless

$\boldsymbol{\varepsilon}^p$ = tensor of plastic strain, dimensionless

$\bar{\varepsilon}_t^p$ = equivalent plastic strain in tension, dimensionless

φ = frictional angle [°]

λ = inelastic multiplier, dimensionless

ν = Poisson's ratio, dimensionless

$\boldsymbol{\sigma}$ = stress tensor, F/L^2 [Pa]

$\bar{\sigma}$ = effective stress, F/L^2 [Pa]

σ = nominal stress, F/L^2 [Pa]

σ_n = the stress in the normal direction to the fracture surface, F/L^2 [Pa]

$\sigma_{n\beta}$ = the normal stress at the surfaces of natural fractures, F/L^2 [Pa]

σ_t and σ_s = stress components in the two tangential directions, F/L^2 [Pa]

σ_1 = the maximum principal stress, F/L^2 [Pa]

σ_3 = the minimum principal stress, F/L^2 [Pa]

σ_{b0} = the strength of bi-axial compression, F/L^2 [Pa]

σ_{c0} = the strength of uniaxial compression, F/L^2 [Pa]

$\Delta\sigma^{ep}$ = the stress increment calculated with elastoplastic damage constitutive relationship, F/L^2 [Pa]

$\Delta\sigma^{\mathrm{r}} =$ the residual stress in an incremental step, F/L^2 [Pa]

$\sigma_{\mathrm{n}} =$ the normal stress at the plane of critical stress status, F/L^2 [Pa]

$\tau =$ the shear stress at the frictional surface, F/L^2 [Pa]

$\tau_\beta =$ the shear stress at the surfaces of natural fractures, F/L^2 [Pa]

$\sigma_{\mathrm{b0}} =$ strength limit of bi-axial compression, F/L^2 [Pa]

$\sigma_{\mathrm{c0}} =$ compressive strength, F/L^2 [Pa]

$\sigma_{\mathrm{cu}} =$ peak value of compressive strength after hardening, F/L^2 [Pa]

$\bar{\sigma}_{\mathrm{c}}(\bar{\varepsilon}_c^{\mathrm{p}}) =$ effective compressive strength, F/L^2 [Pa]

$\sigma_{\mathrm{t0}} =$ tensile strength, F/L^2 [Pa]

$\hat{\sigma}_{\mathrm{t0}} =$ principal stress components, F/L^2 [Pa]

$\hat{\bar{\sigma}}_{\mathrm{max}} =$ maximum principal stress component, F/L^2 [Pa]

$\bar{\sigma}_{\mathrm{t}}(\bar{\varepsilon}_{\mathrm{t}}^{\mathrm{p}}) =$ effective tensile strength, F/L^2 [Pa]

$\sigma_{ij} =$ stress tensor, F/L^2 [Pa]

$\tau =$ shear stress, F/L^2 [Pa]

$\psi =$ fracture density [1/m]

$\omega =$ the continuum-damage variable related to natural fractures, dimensionless

$\boldsymbol{\omega}_{\mathrm{T}} =$ the synthetic damage tensor, dimensionless

$\Psi(\theta, f_i) =$ dilatancy angle [°]

$<.> =$ symbol of Heaviside function, dimensionless

Others

holonomic relationship	the relationship between two variables is given in terms of their total values; an alternative form of relationships to those of incremental form
BHP	bottom hole pressure
FEA	finite element analysis
FG	fracture gradient
GOM	Gulf of Mexico
IPM	injection pressure window
LCM	lost control materials
MWW	mud weight window
OBG	overburden gradient
PP	pore pressure
SRV	stimulated reservoir volume
TVD	true vertical depth

References

Aifantis, E.C. (1992) On the role of gradient in the localization of deformation and fracture. *International Journal of Engineering Science*, 30, 1279–1299.

Aifantis, E.C. (2003) Update on a class of gradient theories. *Mechanics of Materials*, 35:1, 259–280.

Alberty, M.W. & McLean, M.R. (2004) A physical model for stress cages. *SPE Annual Technical Conference and Exhibition, 26–29 September 2004, Houston, TX*. SPE-90493-MS.

Al-saba, M.T., Nygaard, R., Saasen, A. & Nes., O.-M. (2014) Laboratory evaluation of sealing wide fractures using conventional lost circulation materials. *SPE Annual Technical Conference and Exhibition, 27–29 October 2014, Amsterdam, The Netherlands*. SPE-170576-MS.

Alves, M. & Jones, N. (1999) Influence of hydrostatic stress on failure of axisymmetric notched specimens. *Journal of the Mechanics and Physics of Solids*, 47, 643–667.

Bagherian, B., Ghalambor, A., Sarmadivaleh, M., Rasouli, V., Nabipour, A. & Mahmoudi Eshkaftaki, M. (2010) Optimization of multiple-fractured horizontal tight gas well. *SPE International Symposium and Exhibition on Formation Damage Control, 10–12 February 2010, Lafayette, LA*. SPE-127899-MS.

Bahrami, V. & Mortazavi, A. (2008) A numerical investigation of hydraulic fracturing process in oil reservoirs using non-linear fracture mechanics. *5th Asian Rock Mechanics Symposium, 24–26 November 2008, Tehran, Iran*. ISRM-ARMS5-2008-120.

Bartko, K., Al-Shobaili, Y., Gagnard, P., Warlick, M. & Ba Im, A. (2009) Drill cuttings re-injection (CRI) assessment for the Manifa field: an environmentally safe and cost-effective drilling waste management strategy. *SPE Saudi Arabia Section Technical Symposium, 9–11 May 2009, Al-Khobar, Saudi Arabia*. SPE-126077-MS.

Bazant, Z.P. & Cedolin, L. (1991) *Stability of Structures: Elastic, Inelastic, Fracture and Damage Theories*. Oxford University Press, Oxford, NY.

Bazant, Z.P. & Pijaudier-Cabot, G. (1988) Nonlocal continuum damage, localization instability and convergence. *Journal of Applied Mechanics* (ASME), 55, 287–293.

Besson, J. (2001) *Mecanique et Ingenierie des Materiaux – Essais mecaniques, Eprouvettes axisymetriques entaillees (Nonlinear Mechanics of Materials)*. Hermès Science Publishing Ltd, Paris, France (in French).

Bond, A.J., Zhu, D., Kamkom, R. & Hill, A.D. (2006) The effect of well trajectory on horizontal well performance. *International Oil & Gas Conference and Exhibition in China, 5–7 December 2006, Beijing China*. SPE-104183-MS.

Borvik, T., Hopperstad, O.S. & Berstad, T. (2003) On the influence of stress triaxiality and strain rate on the behavior of a structural steel. Part II. Numerical study. *European Journal of Mechanics* A: *Solids*, 22, 15–32.

Buller, D., Hughes, S.N., Market, J., Petre, J.E., Spain, D.R. & Odumosu, T. (2010) Petrophysical evaluation for enhancing hydraulic stimulation in horizontal shale gas wells. *SPE Annual Technical Conference and Exhibition, 19–22 September 2010, Florence, Italy*. SPE-132990-MS.

Capasso, G. & Mantica, S. (2006) Numerical simulation of compaction and subsidence using ABAQUS. *Proceedings of 2006 ABAQUS Users' Conference, Boston, MA*. pp. 125–144.

Carnes, P.S. (1966) Effects of natural fractures or directional permeability in water flooding. *SPE Secondary Recovery Symposium, 2–3 May 1966, Wichita Falls, TX*. SPE-1423-MS.

Caughron, D.E., Renfrow, D.K., Bruton, J.R., Ivan, C.D. Broussard, P.N., Bratton, T.R. & Standifird, W.B. (2002) Unique crosslinking pill in tandem with fracture prediction model cures circulation losses in deepwater Gulf of Mexico. *Presented at the IADC/SPE Drilling Conference, 26–28 February 2002, Dallas, TX*. SPE-74518-MS.

Chaboche, J.L. (1992) Damage induced anisotropy: on the difficulties associated with the active/passive unilateral condition. *International Journal of Damage Mechanics*, 1, 148–171.

Chaboche, J.L. & Cailletaud, G. (1996) Integration methods for complex plastic constitutive equations. *Computer Methods in Applied Mechanics and Engineering*, 133, 125–155.

Chavez, J.C., Gibbons, G.M. & McCurdy, P. (2006) Assuring CRI placement longevity: the role of optimum slurry placement. *SPE/IADC Indian Drilling Technology Conference and Exhibition, 16–18 October 2006, Mumbai, India*. SPE-104047-MS

Christianovich, S.A. & Zheltov, Y.P. (1955) Formation of vertical fractures by means of a highly viscous fluid. *Proceedings 4th World Petroleum Congress 6–15 June, Rome, Italy*, Vol. 2. pp. 579–586.

Cipolla, C.L., Warpinski, N.R., Mayerhofer, M., Lolon, E.P. & Vincent, M.C. (2010) The relationship between fracture complexity, reservoir properties, and fracture-treatment design. SPE *Production and Operations*, 25:04, 438–452. SPE-115769-PA.

Cleary, M.P. (1980) Comprehensive design formulae for hydraulic fracturing. *SPE Annual Technical Conference and Exhibition, 21–24 September 1980, Dallas, TX*. SPE 9259-MS.

Cocchetti, G., Maier, G. & Shen, X.P. (2002) On piecewise linear models of interfaces and mixed mode cohesive cracks. *Journal of Engineering Mechanics* (ASCE), 3:3, 279–298.

Dassault Systèmes (2008) Abaqus analysis user's manual, Vol. 3: Materials, Version 6.8. Dassault Systèmes, Vélizy-Villacoublay, France. 19.3.1–17 – 19.3.2–14.

Dassault Systèmes (2010) Abaqus user's manual, Vol. 2: Analysis, Version 6.10. Dassault Systèms: Vélizy-Villacoublay, France. 10.2.1–1 – 10.2.3–10.

Dassault Systèmes (2011) Abaqus analysis user's manual, Vol. 3: Materials, Version 6.11. Dassault Systemes Simulia Corp., Providence, RI.

de Borst, R., Pamin, J. & Geers, M.G.D. (1999) On coupled gradient-dependent plasticity and damage theories with a view to localization analysis. *European Journal of Mechanics* A: *Solids*, 18, 939–962.

Dean, R.H., Gai, X., Stone, C.M. & Minkoff, S.E. (2003) A comparison of techniques for coupling porous flow and geomechanics. *SPE Reservoir Simulation Symposium, 3–5 February 2003, Houston, TX*. SPE-79709-MS.

Dragon, A. & Mroz, Z. (1979) A continuum model for plastic-brittle behavior of rock and concrete. *International Journal of Mechanical Sciences*, 17, 121–137.

Edwards, S.T., Bratton, T.R. & Standifird, W.B. (2002) Accidental geomechanics – capturing in-situ stress from mud losses encountered while drilling. *SPE/ISRM Rock Mechanics Conference, 20–23 October 2002, Irving, TX*. SPE-78205-MS.

Ehlig-Economides, C.A., Valko, P. & Dyashev, I. (2006) Pressure transient and production data analysis for hydraulic fracture treatment evaluation. *2006 SPE Russian Oil and Gas Technical Conference and Exhibition, 3–6 October 2006, Moscow, Russia*. SPE-101832-RU.

Etse, G. & Willam, K. (1999) Failure analysis of elastoviscoplastic material models. *Journal of Engineering Mechanics* (ASCE), 125:1, 60–69.

Ezell, R., Quinn, F., Chima, J.I. & Baim, A. (2011) First successful field utilization of cuttings re-injection (CRI) in the offshore Manifa field of Saudi Arabia as an environmentally friendly and cost effective waste disposal method. *SPE Annual Technology Conference and Exhibition, 30 October–2 November 2011, Denver, CO*. SPE-147171-MS.

Fatehi, A., Quittmeyer, R., Demirkan, M., Blanco, J. & Kimball, J. (2014) Predicting the seismic hazard due to deep injection well-induced seismicity. *Shale Energy Engineering 2014, 21–23 July 2014, Pittsburgh, PA*. pp. 256–264.

Fett, D., Martin, F., Dardeau, C., Rignol, J., Benaissa, S., Adachi, J. & Pastor, J.A.S.C. (2009) Case history: successful wellbore strengthening approach in a depleted and highly unconsolidated sand in deepwater Gulf of Mexico. *SPE/IADC Drilling Conference and Exhibition, 17–19 March 2009, Amsterdam, The Netherlands*. SPE-119748-MS.

Franquet, J.A., Krisadasima, S., Bal, A. & Pantic, D.M. (2008) Critically-stressed fracture analysis contributes to determining the optimal drilling trajectory in naturally fractured reservoirs. *International Petroleum Technology Conference, 3–5 December 2008, Kuala Lumpur, Malaysia*. IPTC-12669-MS.

Ghavamian, S. & Carol, I. (2003) Benchmarking of concrete cracking constitutive laws: MECA project. In: Bicanic, N., de Borst, R., Mang, H. & Meschke, G. (eds) *Computational Modelling of Concrete Structures*. Swets and Zeitinger, Lisse, The Netherlands. pp. 179–187.

Gopalaratnam, V.S. & Shah, S.P. (1985) Softening response of plain concrete in direct tension. *ACI Journal*, 82:3, 310–323.

Govindjee, S., Kay, G. & Simo, J.C. (1995) Anisotropic modelling and numerical simulation of brittle damage in concrete. *International Journal for Numerical Methods in Engineering*, 38, 3611–3633.

Guenifi, A. (2013) Satellite field exploration in south east of Berkine basin. *Offshore Mediterranean Conference and Exhibition, 20–22 March 2013, Ravenna, Italy*. OMC-2013–019.

Guo, Q., Dutel, L.J., Wheatley, G.B., McLennan, J.D. & Black, A.D. (2000) Assurance increased for drill cuttings re-injection in the Panuke field Canada: case study of improved design. *IADC/SPE Drilling Conference, 23–25 February 2000, New Orleans, LA*. SPE-59118-MS.

Guo, Q., Cook, J., Way, P., Ji, L. & Friedheim, J.E. (2014) A comprehensive experimental study on wellbore strengthening. *IADC/SPE Drilling Conference and Exhibition, 4–6 March 2014, Fort Worth, TX*. SPE-167957-MS.

Gutierrez, M. & Nygard, R. (2008) Shear failure and brittle to ductile transition in shales from p-wave velocity. *42nd US Rock Mechanics Symposium, 29 June–2 July 2008, San Francisco, CA*. ARMA 08–198.

Hanks, T.C. & Kanamori, H. (1979) A moment magnitude scale. *Journal of Geophysical Research*, 84, 2348–2350.

Hashash, Y.M.A., Wotring, D.C., Yao, J.I.C., Lee, J.S. & Fu, Q. (2002) Visual framework for development and use of constitutive models. *International Journal for Numerical and Analytical Methods in Geomechanics*, 26, 1493–1513.

Hillorberg, A., Modeer, M. & Petersson, P.E. (1976) Analysis of crack formation and crack growth in concrete by means of fracture mechanics and finite elements. *Cement and Concrete Research*, 6:6, 773–781.

Himmerlberg, N. & Eckert, A. (2013) Wellbore trajectory planning for complex stress states. *47th U.S. Rock Mechanics/Geomechanics Symposium, 23–26 June, 2013, San Francisco, CA*. ARMA-2013–316.

Horstemeyer, M.F., Lathrop, J., Gokhale, A.M. & Dighe, M. (2000) Modeling stress state dependent damage evolution in a cast Al-Si-Mg aluminium alloy. *Theoretical and Applied Fracture Mechanics*, 33, 31–47.

Kageson-Loe, N.M., Sanders, M.W., Growcock, F., Taugbøl, K., Horsrud, P., Singelstad, A.V. & Omland, T.H. (2009) Particulate-based loss-prevention material – the secrets of fracture sealing revealed! *SPE Drilling and Completion*, 24:04, 581–589.

Kanamori, H. & Brodsky, E. (2001) The physics of earthquakes. *Physics Today*, 4:6, 34–40.

Karsan, I.D. & Jirsa, J.O. (1969) Behavior of concrete under compressive loading. *Journal of Engineering Mechanics* (ASCE), 95:12, 2535–2563.

Karvounis, D., Gischig, V. & Wiemer, S. (2014) Towards a real-time forecast of induced seismicity for enhanced geothermal systems. *Shale Energy Engineering 2014, 21–23 July 2014, Pittsburgh, PA*. pp. 246–255.

Kawamoto, T., Ichikawa, Y. & Kyoya, T. (1988) Deformation and fracturing behavior of discontinuous rock mass and damage mechanics theory. *International Journal for Numerical and Analytical Methods in Geomechanics*, 12, 1–30.

Khennaf, N. & Laddada, A. (2013) Available data and petroleum interest of the Algerian offshore basin. *Offshore Mediterranean Conference and Exhibition, 20–22 March 2013, Ravenna, Italy*. OMC-2013–029.

Khristianovic, S.A. & Zheltov, Y.P. (1955) Formation of vertical fractures by means of a highly viscous fluid. *Proceedings 4th World Petroleum Congress, 6–15 June, Rome, Italy*, Vol. 2. pp. 579–586.

Kristiansen, T.G. (2004) Drilling wellbore stability in the compacting and subsiding Vahall field. *IADC/SPE Drilling Conference, 2–4 March 2004, Dallas, TX*. SPE-87221-MS.

Lee, J. & Fenves, G.L. (1998) A plastic-damage concrete model for earthquake analysis of dams. *Earthquake Engineering and Structural Dynamics*, 27:9, 937–956.

Lemaitre, J. & Chaboche, J.L. (1994) *Mechanics of Solid Materials*. Cambridge University Press, Cambridge, UK.

Lemaitre, J. (1990) *A Course on Damage Mechanics*. 2nd edition, Springer, Berlin, Germany.

Li, Q., Zhang, L. & Ansari, F. (2002) Damage constitutive for high strength concrete in triaxial cyclic compression. *International Journal of Solids and Structures*, 39:15, 4013–4025.

Liebe, T. & Willam, K. (2001) Localization properties of generalized Drucker-Prager elastoplasticity. *Journal of Engineering Mechanics* (ASCE), 127:6, 616–619.

Lietard, O., Unwin, T., Guillot, D. & Hodder, M. (1996) Fracture width LWD and drilling mud / LCM selection guidelines in naturally fractured reservoirs. *European Petroleum Conference, 22–24 October 1996, Milan, Italy*. SPE-36832-MS.

Lubliner, J., Oliver, J., Oller, S. & Onate, E. (1989) A plastic damage model for concrete. *International Journal of Solids and Structures*, 25:3, 299–326.

Manchanda, R., Roussel, N.P. & Sharma, M.M. (2012) Factors influencing fracture trajectories and fracturing pressure data in a horizontal completion. *46th U.S. Rock Mechanics/Geomechanics Symposium, 24–27 June 2012, Chicago, IL*. ARMA-2012–633.

Martin, C.D. & Chandler, N.A. (1993) Stress heterogeneity and geological structures. *International Journal of Rock Mechanics and Mining Sciences*, 30:7, 993–999.

Maxwell, S.C., Jones, M., Parker, R., Miong, S., Leaney, S., Dorval, D., D'Amico, D., Logel, J., Anderson, E. & Hammermaster, K. (2009) Fault activation during hydraulic fracturing. *Proceedings of the 2009 SEG Annual Meeting, 25–30 October 2009, Houston, TX*. SEG-2009–1552.

Mayerhofer, M.J., Lolon, E., Warpinski, N.R., Cipolla, C.L., Walser, D.W. & Rightmire, C.M. (2010) What is stimulated reservoir volume? SPE *Production and Operations*, 23:01, 89–98. SPE-119890-PA.

Mazars, J. & Pijaudier-Cabot, G. (1989) Continuum damage theory – application to concrete. *Journal of Engineering Mechanics* (ASCE), 115:2, 346–365.

McGarr, A. (2014) Maximum magnitude earthquakes induced by fluid injection. *Journal of Geophysical Research: Solid Earth*, 119, 1008–1019.

Mehrabian, A., Jamison, D.E. & Teodorescu, S.G. (2016) Geomechanics of lost-circulation events and wellbore-strengthening operations. *SPE Journal*, 20:06, 1305–1316.

Meschke, G., Lackner, R. & Mang, H.A. (1998) An anisotropic elastoplastic-damage model for plain concrete. *International Journal for Numerical Methods in Engineering*, 42, 703–727.

Miskimins, J.L., Hurley, N.F. & Graves, R.M. (2001) 3D stress characterization from hydraulic fracture and borehole breakout data in a faulted anticline, Wyoming. *SPE Annual Technical Conference and Exhibition, 30 September–3 October 2001, New Orleans, LA*. SPE-71341-MS.

Mortezaei, K. & Vahedifard, F. (2014) Numerical simulation of induced seismicity due to hydraulic fracturing of shale gas reservoirs. *Shale Energy Engineering 2014, 21–23 July 2014, Pittsburgh, PA*. pp. 265–272.

Nagel, N., Zhang, F., Sanchez-Nagel, M., Lee, B. & Agharazi, A. (2013) Stress shadow evaluations for completion design in unconventional plays. *SPE Unconventional Resources Conference Canada, 5–7 November 2013, Calgary, AB, Canada*. SPE-167128-MS.

Narr, W., Schechter, D.W. & Thompson, L.B. (2006) Naturally fractured reservoir characterization. Society of Petroleum Engineers, Richardson, TX.

Nechnech, W., Meftah, F. & Reynouard, J.M. (2002) An elasto-plastic damage model for plain concrete subjected to high temperatures. *Engineering Structures*, 24:5, 597–611.

Owen, D.R.J. & Hinton, E. (1980) *Finite Elements in Plasticity: Theory and Practice*. Pineridge Press Limited, Swansea, UK. pp. 164–179.

Parvizi, H., Rezaei-Gomari, S., Nabhani, F., Turner, A. & Feng, W.C. (2015) Hydraulic fracturing performance evaluation in tight sand gas reservoirs with high perm streaks and natural fractures. *EUROPEC 2015, 1–4 June 2015, Madrid, Spain*. SPE-174338-MS.

Pereira, L.C., Costa, A.M., Sousa, L.C. Jr., Amaral, C.S., Souza, A.L.S., Falcão, F.O.L., Portella, F.A., Silva, L.C.F, Mendes, R.A., Chaves, R.A.P., Roehl, D. & Oliveira, M.F. (2010) Specialist program for injection pressure limits considering fault reactivation criteria. *Proceedings of the 44th U.S. Rock Mechanics Symposium and 5th U.S.-Canada Rock Mechanics Symposium, 27–30 June 2010, Salt Lake City, UT*. ARMA 10–214.

Potluri, N.K., Zhu, D. & Hill, A.D. (2005) The effect of natural fractures on hydraulic fracture propagation. *SPE European Formation Damage Conference, 25–27 May 2005, Sheveningen, The Netherlands*. SPE-94568-MS.

Rafiee, M., Soliman, M.Y., & Pirayesh, E. (2012) Hydraulic fracturing design and optimization: a modification to Zipper Frac. Society of Petroleum Engineers. *SPE Eastern Regional Meeting, 3–5 October 2012, Lexington, KT*. SPE 159786.

Reddoch, J., Taylor, C. & Smith, R. (1996) Successful drill cuttings reinjection (CRI) case history with multiple producing wells on a subsea template utilizing low cost natural oil based mud. *International Petroleum Conference and Exhibition of Mexico, 5–7 March 1996, Villahermosa, Mexico*. SPE-35328-MS.

Rodgerson, J.L. (2000) Impact of natural fractures in hydraulic fracturing of tight gas sands. *SPE Permian Basin Oil and Gas Recovery Conference, 21–23 March 2000, Midland, TX*. SPE-59540-MS.

Saanouni, K., Chaboche, J.L. & Lesne, P.M. (1988) Creep crack growth prediction by a non-local damage formulation. In: Mazars, J. & Bazant, Z.P. (eds) *Proceedings of Europe-US Workshop on Strain Localization and Size Effects in Cracking and Damage, 6–9 September 1988, Cachan, France*. Elsevier Applied Science, London, UK and New York, NY. pp. 404–414.

Saanouni, K., Forster, C. & Hatira, F.B. (1994) On the anelastic flow with damage. *International Journal of Damage Mechanics*, 3, 140–169.

Salehi, S. & Nygaard, R. (2011) Numerical study of fracture initiation, propagation, sealing to enhance wellbore fracture gradient. *45th U.S. Rock Mechanics / Geomechanics Symposium, 26–29 June 2011, San Francisco, CA*. ARMA-11–186.

Salehi, S. & Nygaard, R. (2012) Numerical modeling of induced fracture propagation: a novel approach for lost circulation materials (LCM) design in borehole strengthening applications of deep offshore drilling. *SPE Annual Technical Conference and Exhibition, 8–10 October 2012, San Antonio, TX*. SPE-135155-MS.

Sanad, M., Butler, C., Waheed, A., Engelman, B. & Sweatman, R. (2004) Numerical models help analyze lost-circulation/flow events and frac gradient increase to control an HPHT well in the east Mediterranean Sea. *IADC/SPE Drilling Conference, 2–4 March 2004, Dallas, TX*. SPE-87094-MS.

Savari, S., Whitfill, D.L., Jamison, D.E. & Kumar, A. (2014) A method to evaluate lost-circulation materials – investigation of effective wellbore-strengthening applications. *IADC/SPE Drilling Conference and Exhibition, 4–6 March 2014, Fort Worth, TX*. SPE-167977-MS.

Seweryn, A. & Mroz, Z. (1998) On the criterion of damage evolution for variable multiaxial stress state. *International Journal of Solids and Structures*, 35:14, 1589–1616.

Sfer, D., Carol, I., Gettu, R. & Etse, G. (2002) Study of the behavior of concrete under triaxial compression. *Journal of Engineering Mechanics* (ASCE), 128:2, 156–163.

Shahid, A.S.A., Wassing, B.B.T., Fokker, P.A. & Verga, F. (2015) Natural-fracture reactivation in shale gas reservoir and resulting microseismicity. *Journal of Canadian Petroleum Technology*, 54:06, 450–459, SPE-178437-PA.

Shaoul, J., van Zelm, L. & de Pater, C.J. (2011) Damage mechanisms in unconventional-gas-well stimulation – a new look at an old problem. SPE *Production and Operations*, 26:4, 388–400. SPE-142479-PA.

Shen, X. (2012) Cohesive crack for quasi brittle fracture and numerical simulation of hydraulic fracture. In: Shen, X., Bai, M. & Standifird, W. (eds) *Drilling and Completion in Petroleum Engineering – Theory and Numerical Application*. CRC Press, Boca Raton, FL. pp. 175–179.

Shen, X. (2014) Numerical estimation of casing integrity under injection pressure for fracturing of shale gas formation. *Shale Energy Engineering 2014, 21–23 July 2014, Pittsburgh, PA.* pp 318–325.

Shen, X. & Mroz, Z. (2000) Shear beam model for interface failure under antiplane shear (I)-fundamental behavior. *Applied Mathematics and Mechanics*, English Edition, 21:11, 1221–1228.

Shen, X. & Standifird, W. (2015) Case study on application of cohesive fracture in cuttings reinjection in West Africa. *Proceedings of the 13th ISRM International Symposium on Rock Mechanics, 10–13 May 2015, Montreal, Canada.* ISRM-13CONGRESS-2015–030.

Shen, X., Sikaneta, S. & Ramadhin, J. (2010) Mud weight design for deviated wells in shallow loose sand reservoirs with 3-D FEM. *International Oil and Gas Conference and Exhibition, 8–10 June 2010, Beijing, China.* SPE-130717-MS.

Shen, X.P., Shen, G.X., Chen, L.X. & Yang, L. (2005) Investigation on gradient-dependent nonlocal constitutive models for elasto-plasticity coupled with damage. *Applied Mathematics and Mechanics*, English Edition, 26:2, 218–234.

Skomorowski, N., Dussealut, M.B. & Gracie, R. (2015) The use of multistage hydraulic fracture data to identify stress shadow effects. *49th U.S. Rock Mechanics/Geomechanics Symposium, 28 June–1 July 2015, San Francisco, CA.* ARMA-2015–624.

Smith, M.B. (2010) StimPlan™/InjecPlan™ Version 6.00. NSI Technologies Inc., Tulsa, OK.

Soliman, M.Y., East, L. & Adams, D. (2004) Geo-mechanics aspects of multiple fracturing of horizontal and vertical wells. *SPE International Thermal Operations and Heavy Oil Symposium and Western Regional Meeting, 16–18 March 2004, Bakersfield, CA.* SPE-86992-MS.

Soltanzadeh, H. & Hawkes, C.D. (2007) A semi-analytical model for predicting fault reactivation tendency induced by pore pressure change. *Proceedings of the 1st Canada-U.S. Rock Mechanics Symposium, 27–31 May 2007, Vancouver, Canada.* ARMA-07–202.

Sun, J. & Schechter, D.S. (2014) Optimization-based unstructured meshing algorithms for simulation of hydraulically and naturally fractured reservoirs with variable distribution of fracture aperture, spacing, length and strike. *SPE Annual Technical Conference and Exhibition, 27–29 October 2014, Amsterdam, The Netherlands.* SPE-170703-MS.

Sun, J., Schechter, D. & Huang, C.-K. (2015) Sensitivity analysis of unstructured meshing parameters on production forecast of hydraulically fractured horizontal wells. *Abu Dhabi International Petroleum Exhibition and Conference, 9–12 November 2015, Abu Dhabi, UAE.* SPE-177480-MS.

Swoboda, G., Shen, X.P. & Rosas, L. (1998) Damage model for jointed rock mass and its application to tunneling. *Computer and Geotechnics*, 22:3/4, 183–203.

Teufel, L.W., Hart, C.M. & Sattler, A.R. (1984) Determination of hydraulic fracture azimuth by geophysical, geological, and oriented-core methods at the multiwell experiment site, Rifle, CO. *SPE Annual Technical Conference and Exhibition, 16–19 September 1984, Houston, TX.* SPE-13226-MS.

Verga, F.M., Giglio, G., Masserano, F. & Ruvo, L. (2002) Validation of near-wellbore fracture-network models with MDT. Society of Petroleum Engineers.

Warpinski, N.R. (1985) Measurement of width and pressure in a propagating hydraulic fracture. *SPE Journal*, 25:01, 46–54. SPE-11648-PA.

Warpinski, N.R. (2013) Understanding hydraulic fracture growth, effectiveness, and safety through microseismic monitoring. *ISRM International Conference for Effective and Sustainable Hydraulic Fracturing, 20–22 May 2013, Brisbane, QLD, Australia.* ISRM-ICHF-2013–047.

Wikimedia (2012) 2011 Bohai Bay oil spill. Available from: https://en.wikipedia.org/wiki/2011_Bohai_Bay_oil_spill [accessed October 2016].

Williams, V., McCartney, E. & Nino-Penaloza, A. (2016) Far-field diversion in hydraulic fracturing and acid fracturing: using solid particulates to improve stimulation efficiency. *SPE Asia Pacific Hydraulic Fracturing Conference, 24–26 August 2016, Bejing, China*. SPE-181845-MS.

Willson, S.M., Rylance, M. & Last, N.C. (1993) Fracture mechanics issues relating to cuttings re-injection at shallow depth. *SPE/IADC Drilling Conference, 22–25 February 1993, Amsterdam, The Netherlands*. SPE-25756-MS.

Xu, S.-S., Nieto-Samaniego, A.F. & Alaniz-Álvarez, S.A. (2010) 3D Mohr diagram to explain reactivation of pre-existing planes due to changes in applied stresses. In: Xie, F. (ed) *Rock Stress and Earthquakes*. Taylor and Francis Group, London, UK. pp. 739–745.

Zhang, F., Qiu, K., Yang, X., Burghardt, J., Liu, H., Dong, J. & Luo, F. (2015) A study of the interaction mechanism between hydraulic fractures and natural fractures in the KS tight gas reservoir. *EUROPEC 2015, 1–4 June 2015, Madrid, Spain*. SPE-174384-MS.

Subject index

Multiphysics Modeling

Series Editors: Jochen Bundschuh & Mario César Suárez Arriaga

ISSN:1877-0274

Publisher: CRC/Balkema, Taylor & Francis

1. Numerical Modeling of Coupled Phenomena in Science and Engineering:
   Practical Use and Examples
   Editors: M.C. Suárez Arriaga, J. Bundschuh & F.J. Domínguez-Mota
   2009
   ISBN: 978-0-415-47628-72.

2. Introduction to the Numerical Modeling of Groundwater and Geothermal Systems:
   Fundamentals of Mass, Energy and Solute Transport in Poroelastic Rocks
   J. Bundschuh & M.C. Suárez Arriaga
   2010
   ISBN: 978-0-415-40167-83.

3. Drilling and Completion in Petroleum Engineering: Theory and Numerical Applications
   Editors: Xinpu Shen, Mao Bai & William Standifird
   2011
   ISBN: 978-0-415-66527-8

4. Computational Modeling of Shallow Geothermal Systems
   Rafid Al-Khoury
   2011
   ISBN: 978-0-415-59627-5

5. Geochemical Modeling of Groundwater, Vadose and Geothermal Systems
   Editors: J. Bundschuh & M. Zilberbrand
   2011
   ISBN: 978-0-415-668101-1

6. Mathematical and Numerical Modeling in Porous Media: Applications in Geosciences
   Editors: Martín A. Díaz Viera, Pratap N. Sahay, Manuel Coronado & Arturo Ortiz Tapia
   2012
   ISBN: 978-0-415-66537-7

7. Tubular String Characterization in High Temperature High Pressure Oil and Gas Wells
   J. Xu & Z. Wu
   2015
   ISBN: 978-1-138-02670-4

Printed and bound by CPI Group (UK) Ltd, Croydon, CR0 4YY

24/10/2024

01778290-0004